Design for Vibration Isolation and Impact

A Monograph covering:

- Single Degree Systems
- Forced Vibration & Isolation
- Isolation Examples
- Vibration Absorption
- Impact Loading
- Heavy Springs
- Spring Constants

Carl F. Zorowski
Design Engineering Monograph VII

Design for Vibration Isolation and Impact

Copyright 2017
All Rights Reserved

Design for Vibration Isolation and Impact

Table of Contents

Chapter 1 – Single Degree System

Introduction	1
Idealized One Degree System	1
General Equation of Motion	2
Solution to Differential Equation	3
Initial Conditions of Motion	4
Physical Interpretation	4
Effect of Static Deflection	5
Nonlinearity Representation	6
Changing Initial Conditions	7
Introducing Phase Angle	8
Graphical Solution Presentation	10
Example Application	11
Parallel Spring Constant	12
Series Spring Constant	13

Chapter 2 – Forced Vibration & Isolation

Idealized Model	15
Differential Equation of Motion	16
Complimentary Solution	16
Particular Integral	17
Steady State System Response	18
Maximum Displacement	19

Design for Vibration Isolation and Impact

Maximum Transmitted Force	20
Transmissibility Curve	21
Example Application	22
Harmonic Displacement Input	23
Equation of Motion and Solution	24
Magnification Factor	25
The Effect of Damping	26
Equation of Motion and Solution	27
Magnification Factor	28

Chapter 3 – Isolation Examples

Single Phase Motor Isolation	29
Idealized Dynamics Model	30
Differential Equation of Motion	31
Steady State Solution	32
Maximum Spring Torque	32
Isolation Criteria	34
Spring Process Design	34
Double Harmonic Input	34
Idealized Model	35
Governing Equation and Solution	35
Transmitted Force	36
Resonant Frequencies	37
Graphic Representation	38
Caution	39

Chapter 4 – Vibration Absorption

Introduction	41
Idealized Model	41
Differential Equation of Motion	42
Equations Defining X_1 and X_2	43
X_1 and X_2 Solutions	43
Interpretation of Behavior	44
Resonant Frequencies	45
Graphical Representation	46
Example Application	47
Solution	48

Chapter 5 – Impact Loading

Introduction	55
Idealized Elastic Model	55
Equation of Motion	56
Solution and Initial Conditions	57
Force Following Impact	58
Maximum Impact Force	59
Limiting Case	59
Introducing Drop Height	60
Effect of Height Factor	61
Parametric Study	61
Inelastic Impact	63
Impulse Momentum Model	64
Mass Ratio Effect	66

Modified Impulse Force 67
Nonlinearity of Mass Ratio Effect 68
Velocity Ratio Effect 69
Impulse Momentum Application 69
Graphical Interpretation 70
Kinetic Energy Method 71
Limiting Case Result 72
Example Design Application 73
Variable Impulse Force 74
Time of Event 75

Chapter 6 – Heavy Springs

Introduction 77
Idealized Model 77
Harmonic Response 78
Kinetic Energy of Spring 79
Total Maximum Kinetic Energy 80
Total System Potential Energy 80
Applying Energy Principle 81
Example Problem 82
Include Spring Weight 83
Limiting Case 84
Idealized Wave Mechanics Model 84
Governing Equation and Solution 85
Boundary Conditions 85
Frequency Comparison 86

Spring Constants
    Rods in Extension   89
    Shafts in Torsion   90
    Beams in Bending   90

Design for Vibration Isolation and Impact

Preface

This monograph deals with two important dynamic systems design problems: how to isolate elastically supported bodies from the effects of externally applied periodic excitation and determining equivalent static design loads for bodies that undergo sudden impact.

For simplicity, the model used to investigate the isolation issue is a single degree of freedom system of a mass connected by a linear weightless spring to a fixed base. Movement is restricted to vertical motion. The system is also free of any damping. The external applied periodic disturbance is modeled as a simple harmonic input of finite force or displacement.

The results of these analyses are the maximum values of force or displacement of the body as a function of the driving frequency of the external excitation in steady state motion. This is used to determine a criterion for and the degree of isolation that can be designed into the original system.

Two limiting cases are examined to address the issue of sudden impact. One model consists of elastic bodies that separate following collision with no loss of energy. The second model consists of inelastic bodies that remain fixed to one another following collision in which energy is dissipated. The goal of these analyses is to determine equivalent static design load

Design for Vibration Isolation and Impact

requirements that will result in a level of system robustness to withstand the effects of sudden impact.

Chapter 1 analyzes the free vibration of a single degree of freedom system subjected to two sets of initial conditions that produce motion. The concept of system natural frequency is introduced. Changes in response to the differences in the initial conditions is emphasized.

Chapter 2 investigates how the idealized single degree of freedom system responds to an externally applied steady state harmonic forcing function of alternating force and displacement. The resultant force transmitted to the system base is determined as a function of the frequency of the applied disturbance. The concept of system resonance, transmissibility and magnification are addressed. An isolation criterion is established that defines a range of system natural frequencies that provide reduction of the transmitted force.

Chapter 3 presents the solution to two problems in which the forcing functions differ from the simple harmonic input of Chapter 2. These analyses demonstrate that isolation criteria can be very different and highly dependent on the character of the external disturbance.

Chapter 4 investigates the addition of a second elastically supported mass to the original system. This becomes a two degree of freedom system that possess two distinct natural frequencies and resonance states. Total isolation is achieved for the original mass by the proper selection of the natural frequency of the added mass. This body becomes an absorber of the effects of the external forcing disturbance.

Chapter 5 analyzes the dynamic response of two limiting cases of sudden impact. One is defined by elastic bodies that rebound with no loss of energy following impact. In the second limiting case two inelastic body collide and remain in contact following the event. Energy is dissipated and physical damage generally takes place. These two analyses are used to determine the magnitude of equivalent static loads that become design requirements to insure the physical mechanical strength will withstand the impact.

Chapter 6 develops an approximation for including the mass of the elastic support spring in determining the natural frequency of a single dree of freedom system. The limiting case of the spring as the only mass of the system is checked against an exact wave mechanics analysis.

The numerical solution of an example problem is included in each chapter to demonstrate the application of the subject covered.

The analytical developments used to cover the subject content require the solution of a classic second order ordinary differential equation that governs the response of single degree vibrating system. This is

Design for Vibration Isolation and Impact

covered in detail in the first chapter. The process is repeated as required in subsequent chapters. Its level of mathematical complexity should not be a deterrent to developing an understanding of the subject content. All analytical manipulations are completely discussed.

    This monograph is only a brief overview of what can be a very extensive discourse. It will however provide the reader with a useful introduction and appreciation of vibration isolation and impact.

Carl F. Zorowski
Raleigh, NC
2017

Design for Vibration Isolation and Impact

Design for Vibration Isolation and Impact

Chapter 1 – Single Degree System

Introduction

This chapter deals with the investigation of the dynamic behavior of an idealized single degree of freedom system. Its natural frequency and maximum amplitude of oscillation are determined for two different sets of initial conditions.

Idealized One Degree Model

Shown schematically in Figure 1-1 is the idealized model used in the dynamic analysis of a single degree of freedom system. It consists of a body of weight W suspended by a linear weightless spring from a fixed support. The body is restricted to move only in the x direction under the action of some physical input, the force of gravity and the restraining force of the extended spring. Its displacement will be measured in terms of the vertical coordinate x from the static equilibrium position of the weight.

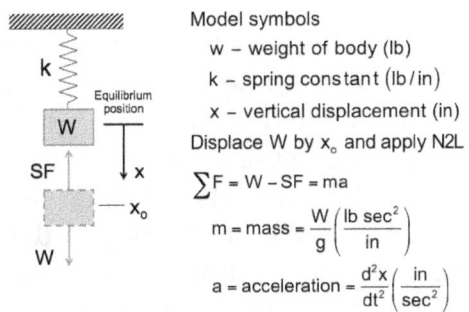

Model symbols
- w – weight of body (lb)
- k – spring constant (lb/in)
- x – vertical displacement (in)

Displace W by x_o and apply N2L

$$\sum F = W - SF = ma$$

$$m = mass = \frac{W}{g}\left(\frac{lb\ sec^2}{in}\right)$$

$$a = acceleration = \frac{d^2x}{dt^2}\left(\frac{in}{sec^2}\right)$$

Figure 1-1 Idealized One Degree Model

Design for Vibration Isolation and Impact

The spring force will be expressed as a constant k times its stretch and t will measure time as the response proceeds.

The analysis is initiated by writing the equation of motion for W assuming it has been displaced a distance x_o and then released. The forces on W will be its weight acting down and the upward tension on the spring, SF.

Applying Newton's second law of motion to this free body the difference in these two forces, W – SF, is set equal to the mass of W times its acceleration. The mass can be expressed as the weight W divided by the gravitation constant g with the acceleration written as the second derivative of x with respect to t, d^2x/dt^2.

General Equation of motion

In Figure 2-2 the spring force, SF, is written as kx + W. The first term comes from the stretch in the spring from x=0 and the second is the weight of the body as supported by the spring in its static equilibrium position.

Substituting the expression for the spring force into the equation of motion permits the weight of the body, W, to be eliminated as a force. This is a consequence of measuring the displacement x from the position of static equilibrium of the weight.

The final result is a second order linear differential equation of a simple oscillator whose solution defines the motion of the body W.

Design for Vibration Isolation and Impact

$$\sum F = ma$$

$$W - SF = \frac{W}{g}\frac{d^2x}{dt^2} \quad \text{but} \quad SF = kx + W$$

$$W - kx - W = \frac{W}{g}\frac{d^2x}{dt^2} \quad \text{or}$$

$$\frac{d^2x}{dt^2} + \frac{kg}{W}x = 0$$

2nd order total differential equation whose solution defines motion as function of time

Figure 1-2   Equation of Motion

Solution to Differential Equation

A solution to the last equation in Figure 1-2 is obtained by recognizing that a function x=f(t) must be such that the second derivative and the function have the same form. Such a function is sin bt or cos bt.

Assume $x = f(t)$ such that $f(t)$ and $\frac{d^2f(t)}{dt^2}$ have the same mathematical form, i.e.

$$x = A\sin bt + B\cos bt$$

now substitute into differential equation

$$-b^2(A\sin bt + B\cos bt) + \frac{kg}{W}(A\sin bt + B\cos bt) = 0$$

or $\left(\frac{kg}{W} - b^2\right)(A\sin bt + B\cos bt) = 0 \Rightarrow b = \sqrt{\frac{kg}{W}}$

then $x = A\sin\omega t + B\cos\omega t$ where $\omega = \sqrt{\frac{kg}{W}}$

Figure 1-3   General Solution for Body Motion

Assuming a general solution of the form   x = A sin bt + B cos bt will satisfy the differential equation if the quantity [(kg/W) − b²] = 0 as shown in Figure 1-3.

Design for Vibration Isolation and Impact

For convenience b is written as $\omega = \sqrt{kg/W}$. The physical significance of ω will be discussed shortly.

Initial Conditions of Motion

The general solution for the equation of motion in Figure 1-3 contains two constants of integration, A and B. These are determined by specifying the initial conditions of motion of the body at t=0. Since the body is displaced a distance x_o and then released with no initial velocity these conditions are expressed as $x=x_o$ and $dx/dt=0$ at t=0. Substituting these conditions into the general solution in Figure 1-4 results in the final expression $x = x_o \cos \omega t$.

Initial conditions of motion
(1) $t = 0 \quad x = x_o$ initial displacement
(2) $t = 0 \quad \dfrac{dx}{dt} = 0$ no initial velocity
Apply (1) to $x = A \sin \omega t + B \cos \omega t$
$x_o = B$
Apply (2) to $\dfrac{dx}{dt} = A\omega \cos \omega t - x_o \omega \sin \omega t$
$0 = A$
so that finally
$x = x_o \cos \omega t$

Figure 1-4 *Applying Initial Conditions of Motion*

Physical Interpretation

Physically this solution indicates that at t=0 the displacement is x_o, where motion begins. With a further increase of t so that $\omega t = \pi$ the displacement is

Design for Vibration Isolation and Impact

$-x_0$ and the stretch in the spring is zero. The motion will then return through x=0 to x=x_0 as ωt proceeds through $3\pi/2$ to 2π. The displacement will then repeat itself as time continues. The body continues to oscillate between x_0 and $-x_0$ as illustrated in Figure 1-5.

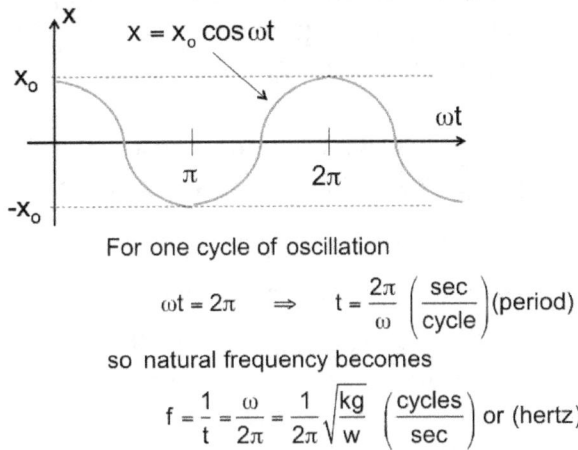

For one cycle of oscillation

$$\omega t = 2\pi \implies t = \frac{2\pi}{\omega} \left(\frac{\sec}{\text{cycle}}\right) \text{(period)}$$

so natural frequency becomes

$$f = \frac{1}{t} = \frac{\omega}{2\pi} = \frac{1}{2\pi}\sqrt{\frac{kg}{w}} \left(\frac{\text{cycles}}{\sec}\right) \text{ or (hertz)}$$

Figure 1-5 Illustrated Oscillatory Motion

The period of this oscillation is determined by setting $\omega t = 2\pi$ and solving for the period as t= $2\pi/\omega$. Frequency is defined as one over the period. The natural frequency of the system becomes f =$\omega/2\pi$ =$(1/2\pi)\sqrt{kg/W}$. Frequency increases with stiffer springs, k, but decreases with increases in weigh, W.

Effect of Static Deflection

It is instructive to look at how the initial stretch, δ, in the spring effects the natural frequency. As shown

Design for Vibration Isolation and Impact

in Figure 1-6 the stretch at static equilibrium of the weight requires that $k\delta = W$. Using this expression to eliminate W from the frequency equation results in f $=(1/2\pi)\sqrt{g/\delta}$. The smaller the initial stretch in the spring the higher the frequency. The change in frequency is non-linear with stretch since it appears in the denominator of the square root term. Higher values of δ effect the frequency less.

Equilibrium position is defined by stretch (δ)
in springs so that spring force balances weight, i.e.

$$k\delta = W \quad \Rightarrow \quad \delta = \frac{W}{k}$$

combine with frequency expression

$$f = \frac{1}{2\pi}\sqrt{\frac{kg}{w}} \quad \text{to become} \quad f = \frac{1}{2\pi}\sqrt{\frac{g}{\delta}}$$

The smaller the initial stretch (δ) in the spring by the weight (w) the higher the natural frequency of the system.

Figure 1-6 *Effect of Stretch, δ, on Frequency*

Nonlinearity Graphical Representation

This non-linearity is illustrated in Figure 1-7 where frequency in cycles per second, cps, or Hertz is plotted as a function of δ in inches. As the static deflection increases the natural frequency decreases and the effect becomes less.

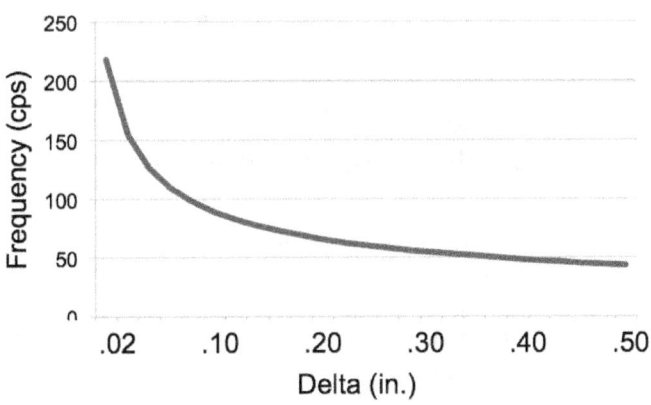

Figure 1-7 Non-linear effect of δ on f

Changing Initial Conditions

Consider how the oscillatory motion changes by a change in the initial conditions. In this example, the body W will be given a positive initial velocity V_o as it is released from it originally displaced position x_0. These new initial conditions are expressed mathematically at the top of Figure 1-8. Since kinetic energy is being added to the potential energy of the displaced body the amplitude of the resulting oscillation will be greater than the previous example.

Substituting these changed initial conditions into the general solution for the motion developed in Figure 1-3 results in finite values for the two constants of integration A and B in Figure 1-8.

Design for Vibration Isolation and Impact

Add initial velocity to displacements, i.e.
(1) $x = x_o \quad t = 0$

(2) $\dfrac{dx}{dt} = v_o \quad t = 0$

Apply (1) to $x = A\sin\omega t + B\cos\omega t$
$x_o = B$

Apply (2) to $\dfrac{dx}{dt} = A\omega\cos\omega t - x_o\omega\sin\omega t$

$v_o = A\omega \quad \Rightarrow \quad A = \dfrac{v_o}{\omega}$
so that finally

$$x = \dfrac{v_o}{\omega}\sin\omega t + x_o\cos\omega t$$

Figure 1-8 Applying New Initial Conditions

This in turn results in a final equation of motion that includes both a sine and cosine term. The maximum deflection of W can be more easily determined by converting this last equation into a single harmonic function by introducing a phase angle into the argument of the trigonometric term.

Introducing Phase Angle φ

This is accomplished in Figure 1-9 by first writing the displacement as $x = X\sin(\omega t + \varphi)$. This is then expanded and matched up with the last equation in Figure 1-8. This gives the result that $X\cos\varphi = V_0/\omega$ and $X\sin\varphi = x_0$. These two expressions are now squared and added together. Recognizing that $\sin^2\varphi + \cos^2\varphi = 1$ results in $X^2 = (V_0/\omega)^2 + x_0^2$.

Design for Vibration Isolation and Impact

Begin with
$$x = X\sin(\omega t + \phi) = X(\sin\omega t \cos\phi + \cos\omega t \sin\phi)$$
Set equal to
$$x = \frac{V_o}{\omega}\sin\omega t + x_o\cos\omega t$$
then $\quad X\cos\phi = \frac{V_o}{\omega} \quad$ and $\quad X\sin\phi = x_o$

square and add
$$X^2(\cos^2\phi + \sin^2\phi) = \left(\frac{V_o}{\omega}\right)^2 + x_o^2$$

Figure 1-9 Introducing a Phase Angle – Step 1

then $\quad X = \sqrt{\left(\frac{V_o}{\omega}\right)^2 + x_o^2}$

and $\quad \frac{X\sin\phi}{X\cos\phi} = \tan\phi = \frac{x_o\omega}{V_o}$

so that finally
$$x = \sqrt{\left(\frac{V_o}{\omega}\right)^2 + x_o^2}\ \sin(\omega t + \phi)$$

and $\quad X_{max} = \sqrt{\left(\frac{V_o}{\omega}\right)^2 + x_o^2}$

Figure 1-10 Introducing a Phase Angle – Step 2

Dividing X $\sin\varphi$ by X $\cos\varphi$ defines the phase angle by the expression $\tan\varphi = x_o\omega/V_o$ as shown in Figure 1-10. In terms of this transformation the displacement as a function of time becomes:

$$x = \sqrt{[(V_o/\omega)^2 + x_o^2]}\ \sin(\omega t + \varphi)$$

9

Design for Vibration Isolation and Impact

Since the maximum value of sin $(\omega t + \varphi)$ is one it is concluded that the maximum displacement of W is simply $x_{max} = \sqrt{[(V_o/\omega)^2 + x_o^2]}$. This is seen to be greater than x_o as was anticipated due to the addition of initial kinetic energy to the system.

Graphical Solution Presentation

A geometric interpretation of this solution for the displacement of W as a function of time in terms of (ωt) is presented in Figure 1-11.

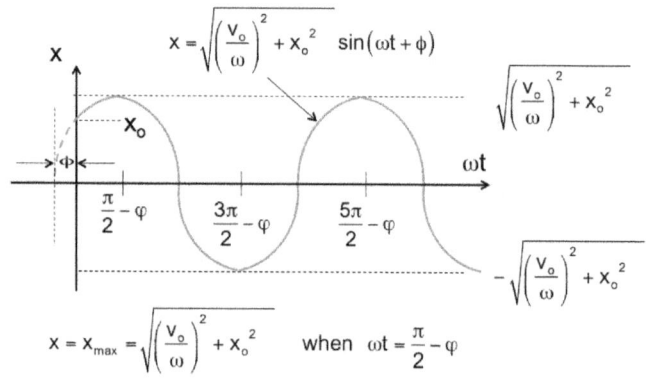

Figure 1-11 Graphical Solution Representation

It is observed in this depiction that the displacement curve begins at $x=x_o$ with a positive slope indicating the initial positive velocity V_o.

The maximum displacement occurs at $\omega t = \pi/2 - \varphi$ and is greater than x_o. The maximum displacement is given by $x_{max} = \sqrt{[(V_o/\omega)^2 + x_o^2]}$.

10

Design for Vibration Isolation and Impact

However, the natural frequency is unaffected since the time for one complete cycle is defined by $\omega t = (\pi/2 - \varphi) + (\varphi)$ and remains the same at $f = (1/2\pi)\sqrt{kg/W}$.

Example Application

A machine tool is supported elastically by a spring system on a fixed base. Its natural frequency is measured to be 5 Hertz (cycles per second). It is proposed to improve its isolation from any input motion from the base by reducing its natural frequency. An additional support system equivalent to that which now supports the machine is available. To reduce the natural frequency of the machine is it better to place the two spring systems in parallel or in series with one another?

The spring constant of the original support arrangement is calculated first. With the natural frequency and the weight known the magnitude of the spring constant can be determined directly from the frequency equation. This calculation is carried out in Figure 1-12.

It is then necessary to determine how this calculated value of the spring constant will be altered by another system placed in parallel or series with it. This is done by analyzing the equilibrium of the two idealized spring models in Figure 1-13 and 1-14.

Design for Vibration Isolation and Impact

Weight of machine — $W = 30000\ lbs$
System natural frequency — $f = 5\ cps\ (hertz)$
$g = 386\ in/sec^2$

begin with

$$f = \frac{1}{2\pi}\sqrt{\frac{kg}{W}}$$

so that $\quad k = \dfrac{4\pi^2 f^2 W}{g}$

then $\quad k = \dfrac{4(3.142)^2(5)^2 3000}{g} = 7663\ lb/in$

Figure 1-12 Original System Spring Constant

The equilibrium of the forces of two springs acting in parallel involving their spring constants and displacements are analyzed in Figure 1-13. If the two springs are equivalent the systems effective spring constant, k_e, is equal to twice the value of the spring constant of each spring, $k_e = 2k$.

Parallel Spring Constant

The sum of the displacements of two springs in series is equal to the displacement of the entire system. This is used to determine the equivalent spring constant of this system in Figure 1-14. The series arrangement for two equal springs results in $k_e = k/2$.

Design for Vibration Isolation and Impact

Springs in Parallel

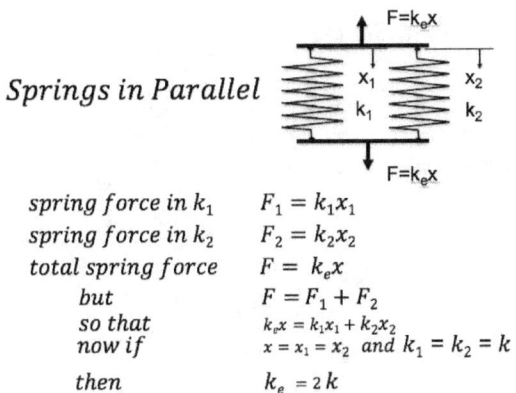

spring force in k_1 $F_1 = k_1 x_1$
spring force in k_2 $F_2 = k_2 x_2$
total spring force $F = k_e x$
but $F = F_1 + F_2$
so that $k_e x = k_1 x_1 + k_2 x_2$
now if $x = x_1 = x_2$ and $k_1 = k_2 = k$
then $k_e = 2k$

Figure 1-13   Equal Springs in Parallel

Writing the frequency of the system with a single spring as $f_o = (1/2\pi)\sqrt{kg/W}$ the modified frequency with the two springs in parallel would be $f_p = \sqrt{2}\, f_o = 1.414 \times 5 = 7.1$ cps.

Series Spring Constant

The spring constant for two springs in series is developed in Figure 1-14. Placing two springs in series as the system support would give $f_s = \sqrt{1/2}\, f_o = 0.71$ or $f_s = 3.6$ cps. Lowering the natural frequency of the machine tool system requires the second equal spring to be placed in series with the original support spring. Whether this provides greater isolation from some input from the base is yet to be determined. This will be investigated in the next chapter.

Design for Vibration Isolation and Impact

Springs in Series

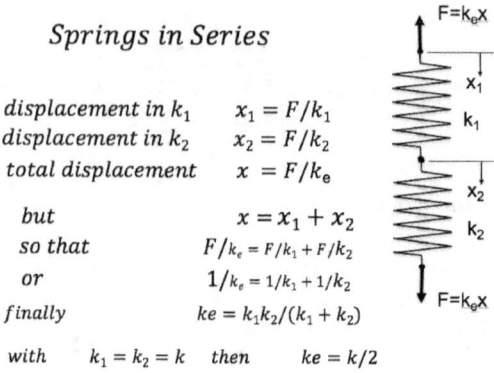

displacement in k_1 $x_1 = F/k_1$
displacement in k_2 $x_2 = F/k_2$
total displacement $x = F/k_e$

but $x = x_1 + x_2$
so that $F/k_e = F/k_1 + F/k_2$
or $1/k_e = 1/k_1 + 1/k_2$
finally $k_e = k_1 k_2 / (k_1 + k_2)$

with $k_1 = k_2 = k$ then $k_e = k/2$

Figure 1-14   Equal Springs in Series

Calculating the magnitude of the original spring system wasn't necessary to answer the posed question of whether a parallel or series arrangement with another equal spring would reduce the frequency. However, it did provide an insight into the high degree of stiffness required for the low system frequency.

Chapter 2 – Forced Vibration & Isolation

The response of a single degree system to an external harmonic excitation is analyzed in this chapter. The resulting behavior is used to define how the system can be isolated from the effects of the input. An over view of the effect of system damping is also presented.

Idealized Model

The model used to analyze how a single degree of freedom system responds to a harmonic input is illustrated in Figure 2-1. A body of weight W is attached by a linear weightless spring to a fixed support. The body is subjected to a harmonic force of the form F sin ωt. Displacement of the body is measured from it static equilibrium position without the forcing function applied. This eliminates the weight, W, as a force in the equation of motion.

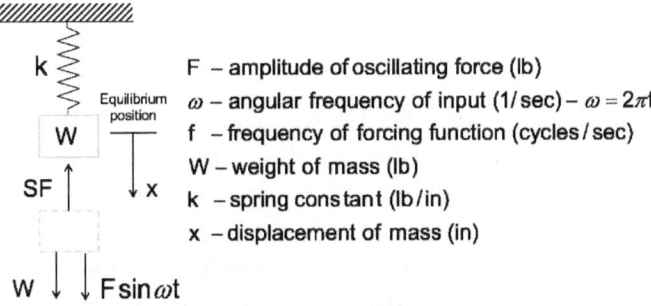

F – amplitude of oscillating force (lb)
ω – angular frequency of input (1/sec) – ω = 2πf
f – frequency of forcing function (cycles/sec)
W – weight of mass (lb)
k – spring constant (lb/in)
x – displacement of mass (in)

Figure 2-1 Idealized Single Degree Model

Design for Vibration Isolation and Impact

Differential Equation of Motion

Newton's Second Law of Motion is now applied to the body in a positive displaced positon. The net forces in the positive x direction will be due to a stretch in the spring, $-kx$, the harmonic forcing function $F \sin \omega t$ and the weight W. Substituting these into N2L and writing the acceleration as d^2x/dt^2 leads to the final equation in Figure 2-2

$$F = ma$$

$$-kx + F \sin \omega t = \frac{W}{g} a$$

$$\text{but } a = \frac{d^2x}{dt^2}$$

therefore

$$\frac{W}{g} \frac{d^2x}{dt^2} + kx = F \sin \omega t$$

$$\text{or } \frac{d^2x}{dt^2} + \frac{kg}{W} x = \frac{Fg}{W} \sin \omega t$$

Figure 2-2   General Equation of motion

Complimentary Solution

Since the right-hand side of this equation is not zero its solution will consist of the sum of two parts, a complimentary solution and a particular integral. The complementary solution, x_c, is obtained by integrating the equation with the right-side set equal to zero. The particular integral is obtained by assuming a form that will satisfy the entire equation. As shown in Figure 2-3 the complimentary solution is developed as it was for the

Design for Vibration Isolation and Impact

single degree of freedom system in Chapter 1. The only change is the designation of the natural frequency of the system, $\omega = \sqrt{kg/W}$ as ω_n to distinguish it from the frequency of the forcing function which was specified as ω.is

assume $x = x_{complementary} + x_{particular}$

x_c is general solution to $\dfrac{d^2 x_c}{dt^2} + \dfrac{kg}{W} x_c = 0$

x_p satisfies the right side of equation

assume $x_c = A\sin bt + B\cos bt$

substituting gives

$-Ab^2 \sin bt - Bb^2 \cos bt + A\dfrac{kg}{W}\sin bt + B\dfrac{kg}{W}\cos bt = 0$

therefore $b^2 = \dfrac{kg}{W} \Rightarrow b = \sqrt{\dfrac{kg}{W}} = \omega_n$

ω_n = natural frequency of spring mass system

Figure 2-3 *Complementary Solution Development*

Particular Integral

It is customary to assume a solution for the particular integral to be of the same mathematical form as the right side of the general equation of motion as presented in Figure 2-4.

This will be successful in this instance as the second derivative with respect to time will still be of the form sin ωt when substituted into the general equation of motion. Substituting this assumed solution into the general equation of motion permits the constant, C, to be uniquely determined.

assume $\quad x_p = C \sin \omega t$

substitute into total differential equation

$$-C\omega^2 \sin \omega t + C\frac{kg}{W} \sin \omega t = F\frac{g}{W} \sin \omega t$$

so $\quad C\left(\dfrac{kg}{W} - \omega^2\right) = F\dfrac{g}{W} \quad$ but $\quad \dfrac{kg}{W} = \omega_n^2$

and $\quad C = \dfrac{F}{k}\left(\dfrac{1}{1-\left(\dfrac{\omega}{\omega_n}\right)^2}\right)$

Figure 2-4   Determining the Particular Integral

Steady State System Response

The final solution for the motion is now the sum of the complementary solution and the particular integral. These are recalled in Figure 2-5.

Final complete solution

$$x = x_c + x_p$$

where $\quad x_c = A \sin \omega_n t + B \cos \omega_n t$

and $\quad x_p = \dfrac{F}{k}\left(\dfrac{1}{1-\left(\dfrac{\omega}{\omega_n}\right)^2}\right) \sin \omega t$

Figure 2-5   Components of Complete Solution

Design for Vibration Isolation and Impact

Although it was assumed that there was no friction in the idealized model in reality there is always some small amount of friction in any real oscillating system. The result of this friction, if it were included in the analysis, would be to completely damp out the complimentary solution in a finite period of time.

This means that x_c can be interpreted as the transient response of the motion leaving x_p as the continuing steady state motion. The steady state motion, x_p becomes the effective long-term motion experienced by the body as a consequence of the harmonic input.

Maximum Displacement

To determine the maximum force experienced by body it is first used necessary to determine the maximum displacement, x_p, of the body relative to its base. This will occur when $\omega t = \pi/2 =$ as indicated in Figure 2-6.

Steady state motion

$$x_p = \frac{F}{k}\left(\frac{1}{1-\left(\omega/\omega_n\right)^2}\right)\sin\omega t$$

Maximum displacement will occur at $\omega t = \pi/2$
when $\sin\omega t = 1$

so that $\quad |x_p|_{max} = \frac{F}{k}\left(\frac{1}{1-\left(\omega/\omega_n\right)^2}\right)$

Figure 2-6   Maximum Steady State Displacement

Design for Vibration Isolation and Impact

Maximum Transmitted Force

The magnitude of the maximum transmitted force, $|F_t|_{max}$, is now determined by multiplying the magnitude of the maximum displacement, $|x_p|_{max}$ by the spring constant k. This is accomplished in Figure 2-7

$$|F_t|_{max} = k|x_p|_{max}$$

$$|F_t|_{max} = F\left(\frac{1}{1-\left(\omega/\omega_n\right)^2}\right)$$

Figure 2-7   Maximum Transmitted Force

The numerical value of the ratio of the maximum force experienced by the body to the amplitude of the forcing function varies dramatically as the ratio the frequency of the forcing function to the natural frequency of the system increases from zero to very high values. This is demonstrated in Figure 2-8.

$$\left(\omega/\omega_n\right) = 0 \quad \left||F_t|_{max}/F\right| = 1$$

$$\left(\omega/\omega_n\right) \Rightarrow 1 \quad \left||F_t|_{max}/F\right| \Rightarrow \infty$$

$$\left(\omega/\omega_n\right) = \sqrt{2} \quad \left||F_t|_{max}/F\right| = 1$$

$$\left(\omega/\omega_n\right) \Rightarrow \infty \quad \left||F_t|_{max}/F\right| \Rightarrow 0$$

Figure 2-8   Effect of $\left(\omega/\omega_n\right)$ on $\left[|F_t|/F\right]$

Design for Vibration Isolation and Impact

As the frequency ratio increases from 0 to 1 the force ratio increases rapidly from 1 to infinity. Then the force ratio returns to 1 at a frequency ratio of $\sqrt{2}$. With very high frequency ratios approaching infinity the force ratio goes to zero.

Transmissibility Curve

This behavior is demonstrated by its graphical representation in Figure *2-9*. When $(\omega/\omega_n) = 1$ the body is said to be in a state of resonance. In reality, this ideal state never exists because of internal damping in the system that absorbs the energy being pumped into it or some physical damage may take place. At $(\omega/\omega_n) = \sqrt{2}$ the force experienced by the body is again equal to the amplitude of the forcing function.

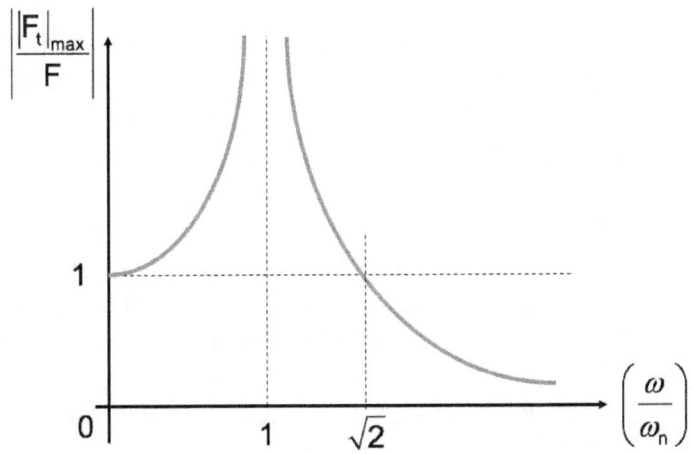

Figure 2-9 Graph of $\left[{}^{[F_t]}\!/_F\right]$ *vs.* (ω/ω_n)

Design for Vibration Isolation and Impact

At higher input frequencies, the body experiences a reduced force from the magnitude of the forcing function. Beyond $(\omega/\omega_n) = \sqrt{2}$ the body experiences an increasing isolation from the effect of the forcing function input. This becomes a design criteria for establishing the required natural frequency of the system to obtain a specified amount of isolation. The behavior of the graph in Figure 2-9 is referred to as the transmissibility effect.

Example Application

The 3000-lb. machine tool in Chapter 1 contains an unbalanced vertical oscillating device that imparts a harmonic forcing function to the machine at the rate of 7 cycles per minute. To what degree, if any, does reducing the natural frequency from 5 cps to 3.5 cps have on the transmissibility of the system?

In Figure 2-10 the transmissibility ratio of the original system with single spring is calculated to be 1.04 indicating that the machine experiences a transmitted force slightly in excess of the magnitude of the forcing function.

By adding a second spring in series with the original support the system natural frequency was lowered to 3.5 cps. The transmissibility ratio at this lower frequency is calculated to be 0.333 in Figure 2-10.

This can be interpreted as providing the revised system with 67% isolation from the effects of the forcing function.

Design for Vibration Isolation and Impact

With $\omega = (2\pi)(7)$ and $\omega_n = (2\pi)(5)$

then $\left(\frac{\omega}{\omega_n}\right)^2 = \left(\frac{7}{5}\right)^2 = 1.96$

and $T.R. = \left|\frac{1}{1-\left(\frac{\omega}{\omega_n}\right)^2}\right| = 1.04 \quad 4\% \quad \text{percent increase}$

With $\omega = (2\pi)(7)$ and $\omega_n = (2\pi)(3.5)$

then $\left(\frac{\omega}{\omega_n}\right)^2 = \left(\frac{7}{3.5}\right)^2 = 4.0$

and $T.R. = \left|\frac{1}{1-\left(\frac{\omega}{\omega_n}\right)^2}\right| = 0.33 \quad 67\% \quad \text{isolation}$

Figure 2-10   Transmissibility Calculations

Harmonic Displacement Input

The response of a single mass system to a harmonic displacement input at the base is now analyzed. The idealized model is shown in Figure 2-11. The coordinate y is introduced to permit the displacement function to be included. Its mathematical form is $y = \delta_o \sin \omega t$.

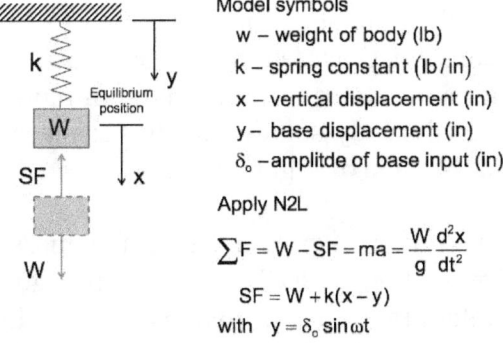

Model symbols
  w – weight of body (lb)
  k – spring constant (lb/in)
  x – vertical displacement (in)
  y – base displacement (in)
  δ_o – amplitde of base input (in)

Apply N2L

$\sum F = W - SF = ma = \frac{W}{g}\frac{d^2x}{dt^2}$

$SF = W + k(x-y)$

with $y = \delta_o \sin \omega t$

Figure 2-11   Harmonic Displacement Input Model

23

Design for Vibration Isolation and Impact

Equation of Motion and Solution

The spring force SF is written as the spring constant, k, times (x - y) plus the weight, W, with x measured from the static equilibrium position of the mass. Substituting this into N2L applied to the mass results in the general differential equation of motion in Figure 2-12.

$$\text{substitute } y = \delta_o \sin \omega t \quad SF = W - k(x - y)$$

$$\text{into } W - SF = \frac{W}{g}\frac{d^2x}{dt^2} \text{ to give}$$

$$\frac{d^2x}{dt^2} + \omega_n^2 x = \omega_n^2 \delta_o \sin \omega t \quad \text{where } \omega_n^2 = \frac{kg}{W}$$

assume for steady state $\quad x = C \sin \omega t$
and substitute into diff. equ.

$$-C\omega^2 \sin \omega t + C\omega_n^2 \sin \omega t = \omega_n^2 \delta_o \sin \omega t$$

$$\text{so} \quad C = \delta_o \frac{1}{1 - \left(\frac{\omega}{\omega_n}\right)^2} \quad \Rightarrow \quad x = \delta_o \frac{1}{1 - \left(\frac{\omega}{\omega_n}\right)^2} \sin \omega t$$

Figure 2-12   Equation of Motion and Solution

The steady state solution is assumed to be $x = C \sin \omega t$. The constant C is determined by substituting the assumed solution into the general equation of motion. This results in the final steady solution for the displacement of the mass in the last line of Figure 2-12.

The maximum displacement of the mass occurs when $\sin \omega t = 1$ at $\omega t = \pi/2$ as shown in Figure 2-13. The dimensionless ratio of x_{max} to δ_o is equal to the same expression involving the forcing frequency and system

Design for Vibration Isolation and Impact

natural frequency as determined in the previous section on transmissibility.

$$x = \delta_o \left(\frac{1}{1-(\omega/\omega_n)^2} \right) \sin \omega t$$

$$x = x_{max} \quad \text{when} \quad \omega t = \frac{\pi}{2} \quad \text{so that}$$

$$\left| \frac{x}{\delta_o} \right|_{max} = \left(\frac{1}{1-(\omega/\omega_n)^2} \right)$$

Figure 2-13   Maximum Displacement

Magnification Factor

The response curve for x_{max}/δ_o as a function of ω/ω_n is shown in Figure 2-14. In this case it is referred to as the magnification factor.

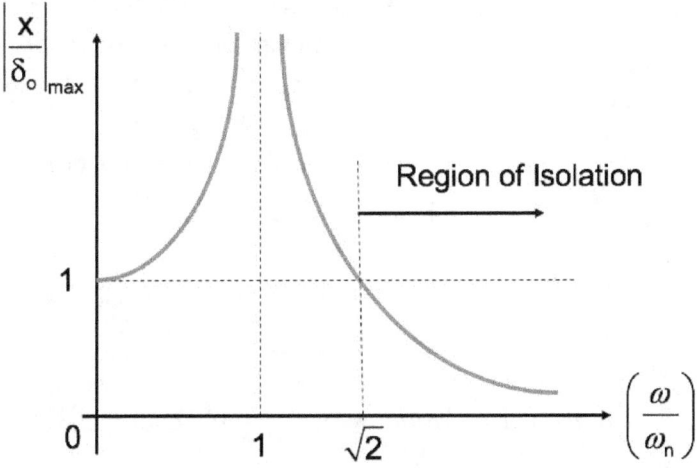

Figure 2-14   Magnification Factor

Design for Vibration Isolation and Impact

The region of isolation from the input forcing function is again achieved and defined by $\left(\frac{\omega}{\omega_o}\right) > \sqrt{2}$.

The Effect of Damping

It was assumed in the previous two analyses that the system was free of any form of friction or other dissipative effects. The analysis is now repeated with viscous damping included. The form of this damping is represented as a restraining force proportional to the velocity of the mass. The model used to include this effect is shown in Figure 2-15 with a viscous damper acting in parallel with the support spring.

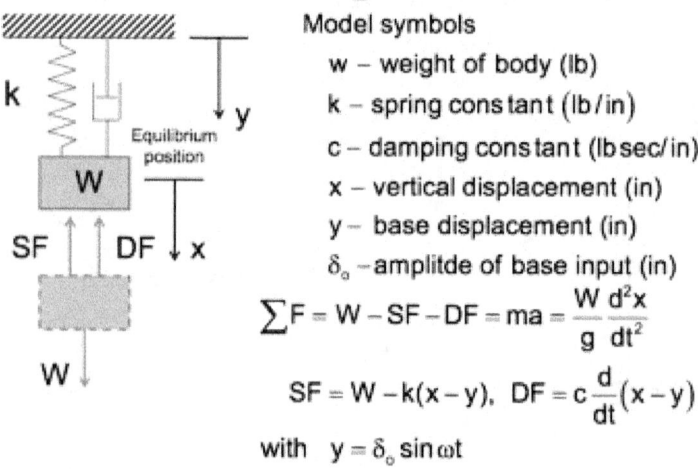

Figure 2-15 Model with Viscous Damper

A harmonic displacement of the base $y = \delta_o \sin \omega t$ will again be used as the forcing function.

Design for Vibration Isolation and Impact

Equation of Motion and Solution

The general differential equation of motion takes the form shown in Figure 2-16. The left side now contains a dx/dt term and the right side has both a sin ωt and cos ωt term. The details of developing the steady state solution for the system motion are not included here for brevity. The final solution is expressed as the ratio, x_{max}/δ_o.

The symbol ξ on the right side is the ratio of the damping constant, c, to the critical damping constant, c_c, where $c_c = 2\sqrt{(kw/g)}$. Critical damping represents that level of damping that is high enough to completely eliminate any harmonic response of the system.

substitute $y = \delta_o \sin\omega t$ $SF = W - k(x-y)$, $DF = -c\dfrac{d}{dt}(x-y)$

into $W - SF - DF = \dfrac{W}{g}\dfrac{d^2x}{dt^2}$ to give

$\dfrac{d^2x}{dt^2} + \left(\dfrac{cg}{W}\right)\dfrac{dx}{dt} + \omega_n^2 x = \delta_o\left(\omega_n^2 \sin\omega t + \left(\dfrac{cg}{W}\right)\omega\cos\omega t\right)$ where $\omega_n^2 = \dfrac{kg}{W}$

solution for Magnification Factor is given by

$\left|\dfrac{x}{\delta_o}\right|_{max} = \dfrac{1}{\sqrt{\left[1-(\omega/\omega_n)^2\right]^2 + \left(2\xi\dfrac{\omega}{\omega_n}\right)^2}}$

where $\xi = \dfrac{c}{c_c}$ with $c_c = 2\sqrt{\dfrac{kW}{g}}$ critical damping constant

Figure 2 – 16   Equation of Motion and Solution

Design for Vibration Isolation and Impact

Magnification Factor

Shown in Figure 2-17 is a graphical representation of the magnification factor with the effect of damping demonstrated.

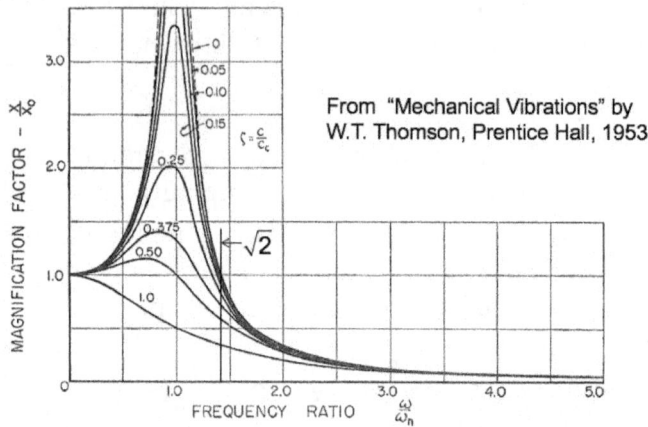

Figure 2-17 Magnification Factor with Damping

Even with a very small value of the damping ratio $\xi = 0.10$ the displacement response remains finite at the frequency ratio that defines resonance. As the damping ratio is increased this maximum displacement continues to decrease and the peak value begins to occur before the resonance frequency is reached.

When $\xi = 1.0$ the maximum value of the magnification factor never exceeds 1.0 and the mass undergoes no harmonic motion.

Design for Vibration Isolation and Impact

Chapter 3 – Isolation Examples

Not all forcing functions are purely harmonic like those already analyzed, i.e. possessing equal positive and negative input amplitudes. In this chapter, the process of analysis in Chapters 1 and 2 is used to investigate the effects on magnification factor and isolation criteria in two instances where the input is not a simple pure harmonic.

Single Phase Motor Isolation

Single-phase ac motors develop a torque that pulsates between zero and twice the average value at double the supply frequency, 120 cycles for a 60-cycle motor. The result is transmission of a pulsating torque through the motor support to its base with potential generation of objectionable noise.

To solve this problem the motor casing is normally supported by rubber rings around the bearing housing to provide the desired torsional springiness for isolation. To avoid the use of rubber in an application of high ozone concentration an all metal mounting is desired.

A proposed design is shown in Figure 3-1. Steel straps bent as indicated and fastened rigidly to the bearing housing and the base will provide both torsional flexibility and lateral stiffness. A suitable isolation criteria is desired for design of the strap dimensions for a 2 horsepower, 60-cycle, 1750 rpm motor.

29

Design for Vibration Isolation and Impact

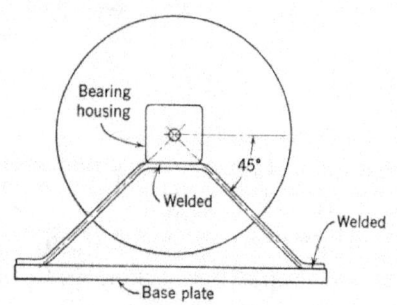

Figure 3-1   Proposed Motor Support

Idealized Dynamic Model

Figure 3-2 shows the idealized model to be used in determining the dynamic response of this system. The angular rotary motion, θ, of the motor casing will be dictated by the combination of the motor driving torque reaction T_m and the restraining torque of the spring support T_s.

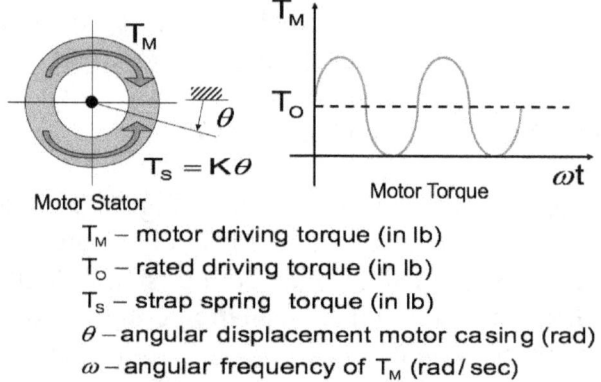

T_M – motor driving torque (in lb)
T_o – rated driving torque (in lb)
T_s – strap spring torque (in lb)
θ – angular displacement motor casing (rad)
ω – angular frequency of T_M (rad/sec)

Figure 3 –2   Idealized Dynamic Model

Design for Vibration Isolation and Impact

Differential Equation of Motion

Applying N2L to define the angular rotary motion of the motor casing results in the motor torque minus the spring torque equal to the rotary inertia of the casing multiplied by the second time derivative of its angular displacement in Figure 3-3. The spring torque is expressed as a rotary spring constant times the angular displacement, $T_s = k\theta$. The motor torque is assumed to be $T_m = T_o(1 - \sin\omega t)$. Substituting these torque expressions into the N2L equation leads to the second order differential equation governng the casing motion in Figure 3-3.

The solution to this equation consists of a complimentary solution and a particular integral. Only the particular integral is of interest as it represents the final steady state rotary behavior of the motor casing.

Apply $\sum T = I\alpha$ to motor casing

$$T_M - T_S = I_o\alpha = I_o \frac{d^2\theta}{dt^2}$$

now $\quad T_S = K\theta$

where K — support strap spring constant (in lb/rad)

and $\quad T_M = T_O(1 + \sin\omega t)$

substituting in equation of motion gives

$$I_o \frac{d^2\theta}{dt^2} + K\theta = T_o(1 + \sin\omega t) \quad \text{or}$$

$$\frac{d^2\theta}{dt^2} + \frac{K}{I_o}\theta = \frac{T_o}{I_o}(1 + \sin\omega t)$$

Figure 3-3   Equation Governing Casing Motion

Design for Vibration Isolation and Impact

Steady State Solution

The steady state motion for θ is assumed to take the form of the right side of the general equation of motion, i.e., $\theta = A + B \sin \omega t$. The constants A and B are determined by substituting this assumed solution into the differential equation of motion. The calculated values of A and B are given in Figure 3-4 where ω_o is the natural frequency of the casing on its spring base, i.e. $\sqrt{(k/I_o)}$.

$$\frac{d^2\theta}{dt^2} + \omega_o^2 \theta = \frac{T_o}{I_o}(1 + \sin \omega t)$$

assume $\theta_p = A + B \sin \omega t$ and substitute into D.E.

$$-B\omega^2 \sin \omega t + \omega_o^2 A + \omega_o^2 B \sin \omega t = \frac{T_o}{I_o} + \frac{T_o}{I_o} \sin \omega t$$

which is satisfied if

$$A = \frac{T_o}{I_o \omega_o^2} = \frac{T_o}{K}$$

$$B = \frac{T_o}{I_o}\left(\frac{1}{\omega_o^2 - \omega^2}\right) = \frac{T_o}{K}\left(\frac{\omega_o^2}{\omega_o^2 - \omega^2}\right) = \frac{T_o}{K}\left(\frac{1}{1-\left(\omega/\omega_o\right)^2}\right)$$

Figure 3 – 4 Constants of the Particular Integral

Maximum Spring Torque

Substituting A and B into the assumed particular integral gives the steady state solution for the angular displacement, θp, in Figure 3-5. The maximum spring torque, T_{smax}, is then obtained by multiplying the steady state solution, θp, by the spring constant, k, to give T_{smax} evaluating θp at $\sin \omega t$ at $\pi/2$. Dividing T_{smax} by T_o results in the final equation for the magnification factor.

Design for Vibration Isolation and Impact

This modified magnification factor is plotted in Figure 3–6 as a function of $(\omega/\omega_o)^2$. The result is dramatically different from that presented on Page 23 even though this is still a ne degree of freedom system.

then $\quad \theta_p = \dfrac{T_o}{K} + \dfrac{T_o}{K}\left(1\Big/\left(1-\left(\omega/\omega_o\right)^2\right)\right)\sin\omega t$

or $\quad \theta_p = \dfrac{T_o}{K}\left[1+\left(1\Big/\left(1-\left(\omega/\omega_o\right)^2\right)\right)\right]\sin\omega t$

maximum support spring torque will occur at $\omega t = \pi/2$ when $\sin\omega t = 1$

$$T_{S_{max}} = \theta_{p_{max}} K = T_o \left[1+\left(1\Big/\left(1-\left(\omega/\omega_o\right)^2\right)\right)\right]$$

or $\quad \dfrac{T_S}{T_{O\,max}} = \left[1+\left(1\Big/\left(1-\left(\omega/\omega_o\right)^2\right)\right)\right]$

Figure 3 – 5   Maximum Spring Torque Ratio

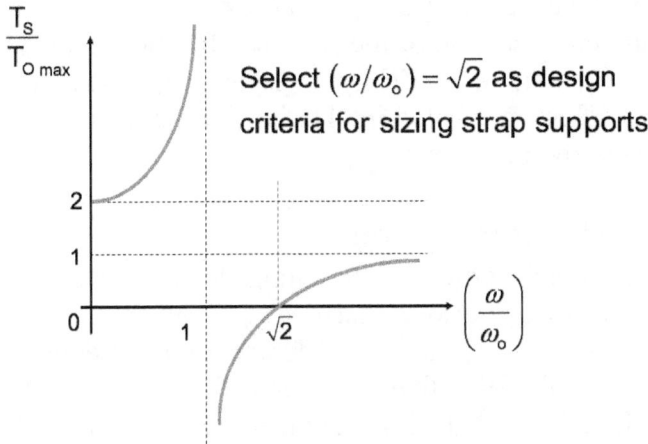

Select $(\omega/\omega_o) = \sqrt{2}$ as design criteria for sizing strap supports

Figure 3 – 6   Modified Magnification Factor

Design for Vibration Isolation and Impact

Isolation Criteria

As seen in this graph the effect of the forcing function can be completely eliminated if (ω/ω_o) is chosen to be $\sqrt{2}$. This becomes the basis for the design of the spring straps that support the motor.

Spring Design Process

The natural frequency of the motor casing on the strap support system is defined as $\omega_o = \sqrt{(k/I_o)}$ where k is the torsional spring constant of the strap support and I_o is the rotary inertia of the motor casing. If the casing is statically rotationally displaced the straps undergo bending and the spring constant k can be related to the length and cross-sectional dimensions of the strap. Then with the isolation criteria defined by $(\omega/\omega_o) = \sqrt{2}$, the driving frequency given as 120 cps and an estimate for Io, the length and cross section of the strap can be chosen to give the required value of k. (This can be accomplished using the methodology presented in "Design for Deflection" but is not be carried out here.)

Double Harmonic Input

A second vertically oscillating device is added to the 3000-lb. machine tool analyzed in Chapter 2. This second harmonic input is of a different forcing frequency and force amplitude to that considered first. The original forcing function, ω_1, had a frequency of 7 cps. The frequency of the second forcing function is 15 cps. The magnitude of the oscillating force of the second forcing function is double that associated with ω_1. Subject to

Design for Vibration Isolation and Impact

these two forcing functions the problem is to specify an isolation criteria for the machine tool support system that will reduce the transmissibility to 1 or less to its base. Will the series spring support that lowered the system natural frequency in Chapter 1 still provide some degree of isolation?

Idealized Model

Illustrated in Figure 3-7 is the idealized model that will be analyzed to determine the transmissibility of system subjected to the double harmonic input.

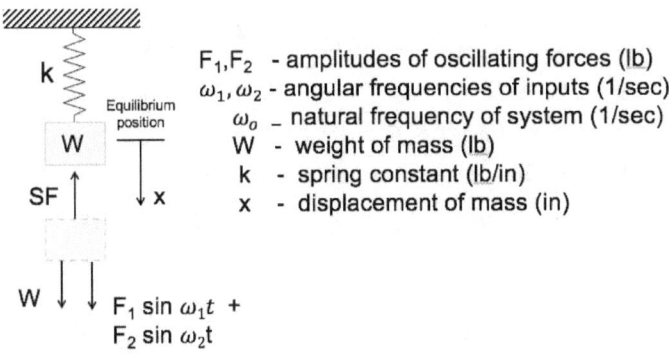

F_1, F_2 - amplitudes of oscillating forces (lb)
ω_1, ω_2 - angular frequencies of inputs (1/sec)
ω_o - natural frequency of system (1/sec)
W - weight of mass (lb)
k - spring constant (lb/in)
x - displacement of mass (in)

Figure 3-7   Idealized Model for Double Harmonic Input

Governing Equation and Solution

The differential equation of motion is obtained by applying NSL to the body in a positive x displacement. This results in the tension in the spring, -kx, plus the two

35

Design for Vibration Isolation and Impact

input forces equal to the mass of the body multiplied by its acceleration shown in Figure 3-8.

Only the steady state solution is desired and is chosen as $x = C_1 \sin \omega_1 t + C_2 \sin \omega_2 t$. The constants C_1 and C_2 are determined by substituting the assumed solution into the differential equation. It is observed that both constants contain the transmissibility function for a simple harmonic input.

$$-kx + F_1 \sin \omega_1 t + F_2 \sin \omega_2 t = \frac{W}{g}\frac{d^2 x}{dt^2}$$

rearranging

$$\frac{d^2 x}{dt^2} + \frac{kg}{W}x = \frac{F_1 g}{W}\sin \omega_1 t + \frac{F_2 g}{W}\sin \omega_2 t$$

assume for steady state

$$x = C_1 \sin \omega_1 t + C_2 \sin \omega_2 t$$

substituting into diff. equ. and simplifying yields

$$C_1 = \frac{F_1}{k}\left(\frac{1}{1-\left(\frac{\omega_1}{\omega_0}\right)^2}\right), \quad C_2 = \frac{F_2}{k}\left(\frac{1}{1-\left(\frac{\omega_2}{\omega_0}\right)^2}\right)$$

Figure 3-8   Governing Equation and Solution

Transmitted Force

The final equation for the displacement x is given in Figure 3-9. The force transmitted to the base is the displacement x multiplied by the spring constant k. By specifying the ratio of F_2 to F_1 as η and the ratio of ω_2 to ω_1 as ε the ratio of F/F_1 can be written in the form of the last equation in Figure 3-9.

Design for Vibration Isolation and Impact

$$x = \frac{F_1}{k}\left(\frac{1}{1-\left(\frac{\omega_1}{\omega_o}\right)^2}\right)\sin\omega_1 t + \frac{F_2}{k}\left(\frac{1}{1-\left(\frac{\omega_2}{\omega_o}\right)^2}\right)\sin\omega_2 t$$

force transmitted to base $F = kx$

let $\frac{F_2}{F_1} = \eta$, $\frac{\omega_2}{\omega_1} = \epsilon$ then

$$\frac{F}{F_1} = \left(\frac{1}{1-\left(\frac{\omega_1}{\omega_o}\right)^2}\right)\sin\omega_1 t + \eta\left(\frac{1}{1-\epsilon^2\left(\frac{\omega_1}{\omega_o}\right)^2}\right)\sin\omega_2 t$$

Figure 3 – 9   Transmitted Force

Resonant Frequencies

Provided ω_1 and ω_2 are not positive multiples of one another there will exist repeated values of time t such that $\sin\omega_1 t$ and $\sin\omega_2 t$ will both be equal to one. This permits the ratio of F_{max}/F_1 to be written as shown in Figure 3–10.

$$\frac{F_{max}}{F_1} = \left|\left(\frac{1}{1-\left(\frac{\omega_1}{\omega_o}\right)^2}\right) + \eta\left(\frac{1}{1-\epsilon^2\left(\frac{\omega_1}{\omega_o}\right)^2}\right)\right|$$

resonance occurs when

$$\left(\frac{\omega_1}{\omega_o}\right)^2 = 1 \text{ and } \left(\frac{\omega_2}{\omega_o}\right)^2 = 1 \text{ or } \epsilon^2\left(\frac{\omega_1}{\omega_o}\right)^2 = 1$$

for the current problem

$f_1 = 7$ cps and $f_2 = 15$ cps   so $\frac{\omega_2}{\omega_1} = \epsilon = \frac{15}{7} = 2.14$

so that for resonance of 15 cps input

$$\left(\frac{\omega_2}{\omega_o}\right)^2 = 1 = (2.14)^2\left(\frac{\omega_1}{\omega_o}\right)^2 \rightarrow \frac{\omega_1}{\omega_o} = \frac{1}{2.14} = 0.46$$

Figure 3 – 10   Resonance Conditions

The right side of this equation indicates that the system can experience two resonance conditions. These will occur when ω_1/ω_o and ω_2/ω_o are each equal to one.

Graphic Representation

Figure 3–11 is a graph of the transmissibility ratio of the machine tool system (the first equation in Figure 3–10) as a function of $\frac{\omega_1}{\omega_o}$. The potential resonance of the second input appears to occur before that of the first input on this plot because its frequency is higher. By setting the transmissibility ratio equal to one it is determined that the magnitude of the maximum transmitted force will occur at values of $\frac{\omega_1}{\omega_o}$ greater than 1.435.

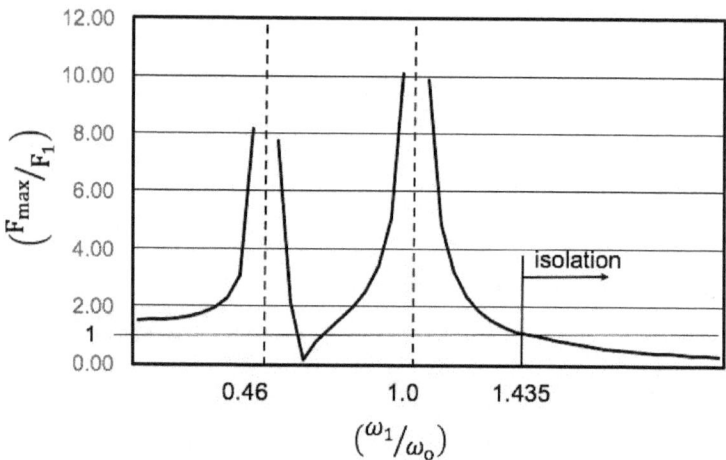

Figure 3-11 Transmissibility Ratio vs. $\frac{\omega_1}{\omega_o}$

Design for Vibration Isolation and Impact

The spring constant associated with the series spring arrangement calculated in Chapter 1 will still provide some degree of isolation even with the second input included. It is left to the reader to show that the degree of insolation in this instance will be 61%

It is also observed that the transmissibility ratio is close to zero at about $\frac{\omega_1}{\omega_o} = 0.55$. This could be used as a design point for establishing the required spring constant for almost complete isolation of the system. However, since it lies between the two resonant frequency ratios it isn't recommended practice unless the two driving frequencies are known to always remain constant.

Caution

These two examples show that the nature of the system support and the form of the external input must be taken into account to properly analyze the response of the system. Isolation as defined by the transmissibility or magnification factor for a single degree system subjected to a single pure harmonic input will not always be directly applicable. The system under consideration together with it support and input must be appropriately idealized before proceeding with determining its steady state response. The resulting transmissibility function can then be analyzed to establish the isolation criteria.

Design for Vibration Isolation and Impact

Design for Vibration Isolation and Impact

Chapter 4 – Vibration Absorption

Introduction

In Chapter 2 a reduction in the effect of an input harmonic disturbance to an elastically supported system was accomplished by properly tuning the natural frequency of the system. This chapter presents a method of reducing the effect of a harmonic input by adding another spring and mass to the original system. This addition absorbs the input energy and requires no change in the natural frequency of the original system.

Idealized model

The model used to examine the process of vibration absorption is shown in Figure 4-1. The Spring k_1 and weight W_1 represent the original system. The weight W_2 and spring k_2 is the added vibration absorber. The dimensions x_1 and x_2 measure the displacement of W_1 and W_2 from their initial positions of static equilibrium. The external harmonic force input, $F_o \sin\omega t$, is applied to W_1.

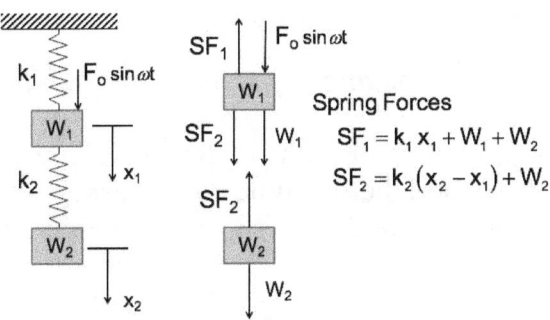

Figure 4-1 Idealized Model

41

Design for Vibration Isolation and Impact

Assuming $x_2 > x_1 > 0$ the forces acting on W1 and W2 will appear as shown in Figure 4-1. The spring force, SF_1 is a result of the extension x_1 and its support of W_1 and W_2 with x_1 measured from the static equilibrium position of W_1. Similarly, SF_2 is due to the differential extension between x_2 and x_1 plus its support of W_2.

Differential Equations of Motion

Applying Newton's 2nd Law of motion to both W_1 and W_2 results in the two differential equations shown in Figure 4-2. The weights W_1 and W_2 do not appear as forces in either equation with x_1 and x_2 measured from the static equilibrium positions of both weights.

Apply NSL to both weights

$$W_1: \quad F_o \sin\omega t - K_1 x_1 + K_2(x_2 - x_1) = \frac{W_1}{g}\frac{d^2 x_1}{dt^2}$$

$$W_2: \quad -K_2(x_2 - x_1) = \frac{W_2}{g}\frac{d^2 x_2}{dt^2}$$

Assume solution to be of the form
$$x_1 = X_1 \sin\omega t$$
$$x_2 = X_2 \sin\omega t$$
and substitute into Diff. Eqs.

Figure 4-2   Equations of Motion for W_1 and W_2

The responses of both weights are assumed to be harmonic with the input disturbance frequency. X_1 and X_2 are the amplitude of these responses, see Figure 4-2.

Design for Vibration Isolation and Impact

Equations Defining X_1 and X_2

The assumed solutions are substituted in to the differential equations of motion. The result is the two simultaneous algebraic equations in Figure 4-3. The natural frequencies of W_1 and W_2 with their respective springs K_1 and K_2 can be written as $\omega_{11} = \sqrt{K_1 g/W_{11}}$ and $\omega_{22} = \sqrt{K_2 g/W_{22}}$. Also, F_o/K_1 can be written as X_o, the static deflection of W_1 at zero input driving frequency. These simplifications result in the two algebraic equations at the bottom of Figure 4-3

Result of substitution

$$\left(K_1 + K_2 - \omega^2 \frac{W_1}{g}\right) X_1 - K_2 X_2 = F_o$$

$$-K_2 X_1 + \left(K_2 - \omega^2 \frac{W_2}{g}\right) X_2 = 0$$

Recognize that $\omega_{11} = \sqrt{K_1 g/W_1}$, $\omega_{22} = \sqrt{K_2 g/W_2}$ and define $X_o = F_o/K_1$

then solution erquations become

$$\left(1 + \frac{K_2}{K_1} - \left(\frac{\omega}{\omega_{11}}\right)^2\right) X_1 - \frac{K_2}{K_1} X_2 = X_o \quad \text{and} \quad -X_1 + \left(1 - \left(\frac{\omega}{\omega_{22}}\right)^2\right) X_2 = 0$$

Figure 4-3 Equations Defining X_1 and X_2

X1 and X2 Solutions

The simultaneous solution of these equations for X_2/X_o and X_1/X_o are presented in Figure 4-4. It is to be noted that these solutions contain terms that are similar to the transmissibility and magnification factors developed in Chapter 2. Interpretation of these results is now required to determine and understand the behavior of this two mass system.

Design for Vibration Isolation and Impact

Solve simultaneously for X_2/X_o and X_1/X_o

$$\frac{X_2}{X_o} = \frac{1}{\left(1 + \frac{K_2}{K_1} - \left(\frac{\omega}{\omega_{11}}\right)^2\right)\left(1 - \left(\frac{\omega}{\omega_{22}}\right)^2\right) - \frac{K_2}{K_1}}$$

and

$$\frac{X_1}{X_o} = \frac{\left(1 - \left(\frac{\omega}{\omega_{22}}\right)^2\right)}{\left(1 + \frac{K_2}{K_1} - \left(\frac{\omega}{\omega_{11}}\right)^2\right)\left(1 - \left(\frac{\omega}{\omega_{22}}\right)^2\right) - \frac{K_2}{K_1}}$$

Figure 4 – 4   Solutions for X_2/X_o and X_1/X_o

This will be done in terms the ratio of the input driving frequency to the natural frequency of W_2, (ω/ω_{22}), recognizing that the added weight, W_2, is expected to provide vibration absorption for W_1.

Interpretation: X_1/X_o and X_2/X_o

Setting $(\omega/\omega_{22}) = 0$, i.e., the driving frequency is zero, then $X_1/X_o = 1$ as indicated in Figure 4-5. This simply states that at zero input driving frequency W_1 undergoes a static deflection due to F_o of magnitude F_o/K_1.

When $(\omega/\omega_{22}) = 1$ the value of $X_1/X_o = 0$ and $X_2/X_o = -F_o/K_2$. The weight W_1 is completely isolated from the harmonic input and W_2 is oscillating with an amplitude equal to its static deflection under the action of F_o. The negative sign indicates that the motion of W_2 is out of phase with the input disturbance.

Design for Vibration Isolation and Impact

If $\dfrac{\omega}{\omega_{22}} = 0$ then $\dfrac{X_1}{X_o} = 1$

If $\dfrac{\omega}{\omega_{22}} = 1$ then $\dfrac{X_1}{X_o} = 0$ and $\dfrac{X_2}{X_o} = -\dfrac{K_1}{K_2} \Rightarrow X_2 = -\dfrac{F_o}{K_2}$

Setting the denominator of either $\dfrac{X_1}{X_o}$ or $\dfrac{X_2}{X_o}$ equal to 0 will give two values of $\dfrac{\omega}{\omega_{22}}$ representing resonant frequencies

Figure 4 – 5 Behavior of X_2/X_o and X_1/X_o

If the added weight, W_2, with its spring support, K_2 is selected such that its natural frequency is the same as the frequency of the input harmonic then the original system, W_1, together with K_1 will be completely isolated from the effects of the external disturbance. This is referred to as "tuning" the vibration absorber.

In Chapter 2 isolation was obtained by decreasing the natural frequency of the system relative to the frequency of the external input. For a simple spring weight system subjected to a pure harmonic input this would require an infinitely soft spring to obtain complete oscillation. One of the advantages of adding a tuned vibration absorber is that the original system requires no change.

Resonant Frequencies

It is also observed that if the denominators of X_2/X_o and X_1/X_o in Figure 4-5 are zero then these ratios become infinite indicating resonance takes place.

Design for Vibration Isolation and Impact

Set $\left(1+\dfrac{K_2}{K_1}-\left(\dfrac{\omega}{\omega_{11}}\right)^2\right)\left(1-\left(\dfrac{\omega}{\omega_{22}}\right)^2\right)-\dfrac{K_2}{K_1}=0$

Replace $\dfrac{K_2}{K_1}$ by $\left(\dfrac{W_2}{W_1}\right)\left(\dfrac{K_2 g}{W_2}\right)\left(\dfrac{W_1}{K_1 g}\right)=\left(\dfrac{W_2}{W_1}\right)\left(\dfrac{\omega_{22}}{\omega_{11}}\right)^2$

and set $\left(\dfrac{W_2}{W_1}\right)=\mu$

then the resonant frequency equation becomes

$\left(\dfrac{\omega_{22}}{\omega_{11}}\right)^2 \left(\dfrac{\omega}{\omega_{22}}\right)^4 - \left(1+(1+\mu)\left(\dfrac{\omega_{22}}{\omega_{11}}\right)^2\right)\left(\dfrac{\omega}{\omega_{22}}\right)^2 + 1 = 0$

to give solutions $\left(\dfrac{\omega_{S1}}{\omega_{22}}\right)$ and $\left(\dfrac{\omega_{S2}}{\omega_{22}}\right)$

Figure 4-6. Resonant Frequency Equation

To determine the two natural frequencies, ω_{S2} and ω_{2S1}, of the combined system that define these resonant conditions requires solving the resulting quadratic equation for the two roots of $(\omega/\omega_{22})^2$. The development of the system resonant frequency equation is presented in Figure 4-6.

Graphical Representation

The point where $\omega/\omega_{22} = 1$ lies on a plot of X_1/X_0 in terms of ω/ω_{22} between ω_{S1}/ω_{22} and ω_{S2}/ω_{22} as illustrated in Figure 4-7.

The importance of knowing the resonant frequencies is that if ω/ω_{22} passes through these frequencies in some transient behavior prior to settling at $\omega/\omega_{22} = 1$ then the W_1-K_1 system must be able to

Design for Vibration Isolation and Impact

withstand the elevated values of X_1/X_o it may be subjected to.

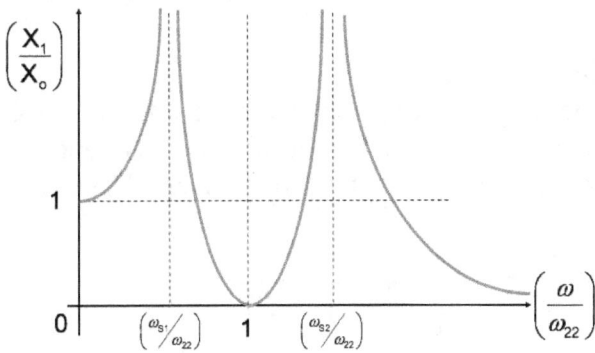

Figure 4 – 7 Behavior of X_1/X_o with ω/ω_{22}

Example Application

A small instrument lathe weighing 350 lbs. is mounted on a base that is subjected to a steady harmonic input of 0.01 inches at a frequency of 10 cps. The natural frequency of the lathe system is measured to be 15 cps. To insure the operational accuracy of the lathe it is recommended that its vibrational response to any steady displacement input be no more than 0.0005 inch in amplitude. The following questions have been raised.

1. What will be the response of the lathe as installed?
2. If the recommended response is exceeded how should the spring constant of its support system be modified?

3. If a vibration absorber is employed what are the design parameters associated with the addition of a 50-lb. weight to the system?
4. What would you recommend as a solution?

Solution

The magnification factor for the original system is calculated in Figure 4-8 for the driving frequency of 10 cps. and original system natural frequency of 15 cps.

$$\frac{X}{X_o} = \frac{1}{1-\left(\frac{\omega}{\omega_o}\right)^2} = \frac{1}{1-\left(\frac{10}{15}\right)^2} = 2.9$$

$X = (2.9) X_o = (2.9)(0.01) = 0.03$ in.

Does not satisfy criteria, i.e $0.03 > 0.0005$ in.

Figure 4 – 8   Original System Magnification Factor

The calculated steady state amplitude, X, of the system of 0.03 in. does not satisfy the specified criteria for the allowable system maximum response.

The natural frequency required of the system to meet the specified isolation response criteria of 0.0005" is determined in Figure 4-9. The magnification factor is set equal to 0.05 and the natural frequency required of the system to achieve this level is determined to be 2.8 cps. With the driving frequency 10 cps the required system natural frequency is 2.18 cps.

Design for Vibration Isolation and Impact

The questions that now arise are:
1. What should the equivalent spring constant be for this revised system?
2. Can this be achieved by adding a spring in series or parallel with the existing system support and what should its spring constant be?
3. Does this represent a mechanically feasible solution?

To satisfy isolation criteria

$$\frac{X}{X_o} = \frac{0.0005}{0.01} = 0.05$$

then $\dfrac{1}{1-\left(\frac{\omega}{\omega_o}\right)^2} = 0.05$

from which

$$\left(\frac{\omega}{\omega_o}\right)^2 = 21 \quad \rightarrow \quad \omega_o = \frac{10}{4.58} = 2.18\ cps$$

Figure 4-9 Natural Frequency for Isolation

The equivalent spring constant required to give the system a natural frequency of 2.8 cps is calculated in Figure 4-10.

Lowering the natural frequency of the system takes a lower equivalent spring constant. This is achieved by placing a softer spring in series with the original system support.

Design for Vibration Isolation and Impact

System Equivalent Spring Constant

$$\omega = \sqrt{\frac{kg}{W}} \quad \rightarrow \quad k = \frac{W\omega^2}{g}$$

$$\omega^2 = (2.8)^2 \times (2\pi)^2 = 187 \ rad^2/sec^2$$

$$k = \frac{(350)(187)}{386} = 170 \ lb/in$$

Figure 4 – 10   *System Equivalent Spring Constant*

The value of this softer spring is now determined in Figure 4–11 where k_e, k_1 and k_2 refer to the equivalent spring constant, the original system spring constant and the spring constant of the spring added in series.

$$k_e = \frac{k_1 k_2}{k_1 + k_2} \quad \rightarrow \quad k_2 = \frac{k_1 k_e}{k_1 - k_e}$$

$$k_1 = \frac{W\omega_1^2}{g} = \frac{(350)(2\pi \times 15)^2}{(386)} = 8046 \ lb/in$$

$$k_e = 170 \ lb/in$$

$$k_2 = \frac{(8046)(170)}{(8046) - (170)} = 174 \ lb/in$$

Figure 4 – 11   *Value of K for Added Spring*

The spring support of the original system has little effect on the added spring. An added design consideration is the magnitude of the initial static deflection of the combined springs. This is calculated as the sum of the deflections of each spring subjected to the total weight of the system as illustrated in Figure 4-12.

Design for Vibration Isolation and Impact

$$\delta_T = \delta_1 + \delta_2$$

$$\delta_T = \frac{W}{k_1} + \frac{W}{k_2}$$

$$\delta_T = \frac{350\ lb}{8046\ lb/in} + \frac{350\ lb}{174\ lb/in}$$

$$\delta_T = 0.043 + 2.01 = 2.053\ inch$$

Figure 4 -12   Total Static Deflection

A total static deflection of 2 inches seems feasible as a design requirement for the added spring.

Now consider the effect of adding a 50-lb. vibration absorber to satisfy the isolation criteria specified for the lathe. The advantage of this solution is that by tuning the add weight to the driving frequency of the external input the response of the lathe will be zero. To determine the amplitude of the steady state response of the added weight it is first necessary to determine its required spring constant, k_2. In Figure 4-13 this is calculated to be 3575. The static displacement design requirement for the added weight will be of no real consequence at (350 lb.)/(3776 lb./in) = 0.10.

$$\omega_{22} = \sqrt{\frac{k_2 g}{W_2}} \rightarrow k_2 = \frac{W_2 \omega_{22}^2}{g}$$

$$\omega_{22}^2 = (10)^2 \times (2\pi)^2 = 3944\ rad^2/sec^2$$

$$k_2 = \frac{(350)(3944)}{386} = 3576\ lb/in$$

Figure 4 – 13   Spring Constant of Added Weigh

Design for Vibration Isolation and Impact

With the spring constants for both the original system and the added weight known the analysis from Figure 4-5 can be used to determine the steady state amplitude response of the vibration absorber, i.e.

The response amplitude of the absorber calculated in Figure 4-14 is certainly of acceptable magnitude for a frequency of 10 cps. If the weight were reduced to 25 lbs. the response amplitude would double, still an acceptable magnitude.

$$\text{if } \left(\frac{\omega}{\omega_{22}}\right) = 1 \quad (a \text{ tunned absorber})$$

$$\text{then } \left(\frac{X_1}{X_o}\right) = 0 \quad \text{and} \quad \left(\frac{X_2}{X_o}\right) = \frac{k_1}{k_2}$$

$$\left(\frac{X_2}{X_o}\right) = \frac{8046}{3576} = 2.25$$

so $X_2 = (2.25)(0.01) = .023$ inch

Figure 4 – 14 *Amplitude of Absorber Weight*

It would appear that using a vibration absorber to satisfy the prescribed isolation requirement for the lathe would be the better choice provided no physical design issues currently unforeseen exist.

The resonance frequencies that straddle the driving frequency are calculated in Figure 4-15 as a matter of curiosity.

Design for Vibration Isolation and Impact

$$\left(\frac{\omega_{22}}{\omega_{11}}\right)^2 \left(\frac{\omega}{\omega_{22}}\right)^4 - \left[1 + (1+\eta)\left(\frac{\omega_{22}}{\omega_{11}}\right)^2\right]\left(\frac{\omega}{\omega_{22}}\right)^2 + 1 = 0$$

$$\left(\frac{10}{15}\right)^2 \left(\frac{\omega}{\omega_{22}}\right)^4 - \left[1 + (1+\frac{50}{350})\left(\frac{10}{15}\right)^2\right]\left(\frac{\omega}{\omega_{22}}\right)^2 + 1 = 0$$

$$\left(\frac{\omega}{\omega_{22}}\right)^4 - 3.41\left(\frac{\omega}{\omega_{22}}\right)^2 + 2.27 = 0$$

$$\left(\frac{\omega}{\omega_{22}}\right)^2 = 0.93, 2.43 \quad \rightarrow \quad \left(\frac{\omega}{\omega_{22}}\right) = 0.96, 1.57$$

$\omega_{s1} = 9.6$ cps $\quad \omega_{s2} = 15.7$ cps

Figure 4-15   Resonant Frequencies

It is interesting to note how close these resonant frequencies of the system are to the driving frequency and the natural frequency of the original system. The external excitation frequency will probably transition from zero to its final value and the system will pass through the resonance value of 9.5 cps.   If this occurs quite quickly and is not repetitive then the natural damping in the system should keep the original system from undergoing any excessive response. However, the absorber must be carefully tuned since there is very little margin between 10 cps and the first system resonance.

Design for Vibration Isolation and Impact

Design for Vibration Isolation and Impact

Chapter 5 – Impact Loading

Introduction

This chapter deals with the dynamics of physical impacts, the response of bodies that suddenly collide with one another. Two limiting cases are investigated: elastic impact in which bodies rebound following sudden contact with no loss in energy and inelastic impact after which bodies remain in contact with one another and energy is dissipated. Methods are presented for estimating equivalent static loadings that can be used as design criteria requirements in these two instances.

Idealized Elastic Model

The idealized model for analyzing the response of elastic bodies colliding is presented in Figure 5-1. The elasticity of a falling weight, W, is represented by the linear spring k_1.

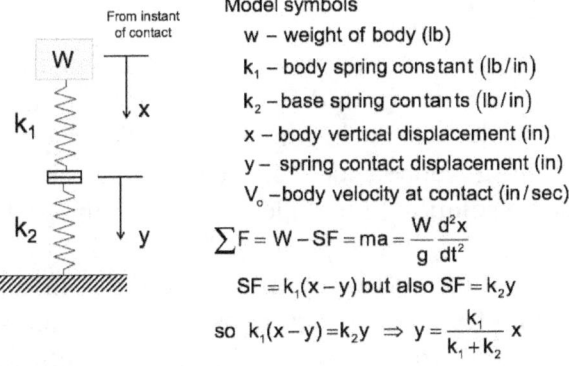

Model symbols
w – weight of body (lb)
k_1 – body spring constant (lb/in)
k_2 – base spring contants (lb/in)
x – body vertical displacement (in)
y – spring contact displacement (in)
V_o – body velocity at contact (in/sec)

$$\sum F = W - SF = ma = \frac{W}{g}\frac{d^2x}{dt^2}$$

$SF = k_1(x-y)$ but also $SF = k_2 y$

so $k_1(x-y) = k_2 y \Rightarrow y = \frac{k_1}{k_1 + k_2} x$

Figure 5-1 Idealized Elastic Model

Design for Vibration Isolation and Impact

The fixed base onto which W is dropped possesses elasticity represented by k_2. The coordinate x measures the displacement of W from the instant that contact takes place. The displacement of the point of contact at the ends of the two springs following contact is measured by the coordinate y.

The two contact pads at the ends of the springs are considered massless eliminating a two degree of freedom problem. The springs are free from any deformation prior to impact. V_o is the vertical velocity of W at the instance of impact.

Newton's second law is applied to the body following impact to obtain the equation governing its dynamic response. The two forces acting on the body are its weight and the spring force of k_1. Recognizing that the force in both springs is the same leads to the relationship between x and y shown in Figure 5-1.

Equation of Motion

Eliminating "y" in terms of "x" in the equation of motion results in the final differential equation in the displacement of the weight in Figure 5-2. The right side of the equation, g, represents a constant forcing function that requires the solution to include a particular integral.

The effect of the two springs, k_1 and k2, in series appears as an effective spring constant, k_e. This will dictate the natural frequency of the system following impact.

Design for Vibration Isolation and Impact

with $y = \dfrac{k_1}{k_1+k_2} x$ then

$$W - SF = W - k_1(x-y) = W - k_1 x\left(1 - \dfrac{k_1}{k_1+k_2}\right) = \dfrac{W}{g}\dfrac{d^2x}{dt^2}$$

which becomes

$$W - x\left(\dfrac{k_1 k_2}{k_1+k_2}\right) = \dfrac{W}{g}\dfrac{d^2x}{dt^2}$$

and finally

$$\dfrac{d^2x}{dt^2} + \dfrac{k_e g}{W} x = g$$

where $k_e = \left(\dfrac{k_1 K_2}{k_1+k_2}\right)$ equivalent spring constant

Figure 5 – 2   Final Equation of motion

Solution and Initial Conditions

The solution for the displacement following impact consists of the complimentary and particular integrals shown in Figure 5-3. The complementary solution will be harmonic with the natural frequency defined by the effective spring constant, k_e.

$$x_c = A\sin\omega t + B\cos\omega t \quad \text{with} \quad \omega = \sqrt{\dfrac{k_e g}{W}}$$

$$x_p = C \text{ (const)} \Rightarrow C = \dfrac{W}{k_e} \quad (lb)\left(\dfrac{in}{lb}\right) = (in)$$

Initial conditions of motion

at $t = 0 \quad x = 0 \quad \dfrac{dx}{dt} = V_o \quad$ then

$$B = -\dfrac{W}{k_e} \quad \text{and} \quad A = \dfrac{\omega}{V_o}$$

Figure 5 – 3   Solution and Initial Conditions

Design for Vibration Isolation and Impact

The particular integral will simply be a constant, $C = W/k_e$. At the instant of impact the displacement of the weight will be zero but its velocity will be V_o. These conditions require the two constants in the complementary solution to be:

$$A = \frac{\omega}{V_o} \text{ and } B = -\frac{W}{k_e}..$$

Force Following Impact

With the displacement of the weight defined the force it experiences following impact is given by the displacement multiplied by the effective spring constant. This is presented in Figure 5-4 in the dimensionless form F/W.

$$x = \frac{V_o}{\omega}\sin\omega t - \frac{W}{k_e}\cos\omega t + \frac{W}{k_e}$$

Force experienced by body (W)

$$F = k_e x = k_e \left(\frac{V_o}{\omega}\sin\omega t - \frac{W}{k_e}\cos\omega t + \frac{W}{k_e} \right)$$

or

$$\frac{F}{W} = \frac{V_o k_e}{\omega W}\sin\omega t - \cos\omega t + 1$$

Now transform into single trig function to determine maximum value of $\frac{F}{W}$

Figure 5-4   Dimensionless Impact Force

The equation for F/W is now transformed into a single trigonometric function with a phase angle and examined for its maximum value. This magnitude

Design for Vibration Isolation and Impact

becomes the effective static force the dropped body should be designed to withstand.

Maximum Impact Force

The determination of the maximum impact force is presented in Figure 5-5. The equation for the dimensionless force is first transformed into a single harmonic function. This transformation is maximized when $(\omega t + \phi = \pi/2)$. As anticipated the force is dependent on the velocity of impact.

$$\frac{F}{W} = \frac{V_o k_e}{\omega W} \sin \omega t - \cos \omega t + 1$$

Assume

$$\frac{F}{W} = D \sin(\omega t + \phi) + 1 \quad \text{where}$$

$$D = \sqrt{1 + \left(\frac{V_o k_e}{\omega W}\right)^2} \quad \text{and} \quad \tan \phi = \frac{\omega W}{V_o k_e} \left(\frac{\text{rad}}{\text{sec}}\right)(\text{lb})\left(\frac{\text{sec}}{\text{in}}\right)\left(\frac{\text{in}}{\text{lb}}\right)$$

maximum force will occur when $(\omega t + \phi) = \pi/2$ giving

$$\left|\frac{F}{W}\right|_{max} = 1 + \sqrt{1 + \left(\frac{V_o k_e}{\omega W}\right)^2}$$

Figure 5 – 5 Maximum Impact Force

Limiting Case

An interesting limiting case is when the height through which the weight is dropped is zero. In this instance the impact velocity, V_o, is zero. The maximum impact force is seen to be twice the weight of the dropped body, see Figure 5-6. This is a frequently referenced result and has been used by some designers to double the factor of safety when impact is anticipated.

Design for Vibration Isolation and Impact

However, this simplification is no longer necessary as a more accurate equivalent static design force can now be determined.

$$\left|\frac{F}{W}\right|_{max} = 1 + \sqrt{1 + \left(\frac{V_o k_e}{\omega W}\right)^2}$$

If $V_o = 0$, i.e. body is droped through zero height (h) then

$$\left|\frac{F}{W}\right|_{max} = 1 + 1 = 2 \quad \text{classic referenced result}$$

Figure 5 – 6   Limiting Case of $V_o = 0$

Introducing Drop Height

The impact velocity can be replaced by $V_o = \sqrt{2gh}$ for a body that free falls vertically through a height h prior to impact. If the natural frequency is also introduced in terms of system parameters the maximum force takes the form of the last equation in Figure 5-7.

$$\left|\frac{F}{W}\right|_{max} = 1 + \sqrt{1 + \left(\frac{V_o k_e}{\omega W}\right)^2}$$

Consider general case where $h \neq 0$

with $\omega^2 = \dfrac{k_e g}{W}$ and $V_0 = \sqrt{2gh}$ free fall of body (W)   then

$$\left(\frac{V_o k_e}{\omega W}\right)^2 = (2gh)\left(\frac{W}{k_e g}\right)\left(\frac{k_e}{W}\right)^2 = 2\frac{k_e h}{W}$$

$$\left|\frac{F}{W}\right|_{max} = 1 + \sqrt{1 + 2\frac{k_e h}{W}}$$

Figure 5 – 7   Introduction of Drop Height

Design for Vibration Isolation and Impact

Effect of Height Factor

The effect of the free fall height factor is demonstrated graphically in Figure 5-8. When $(K_e h/W)$ is zero the maximum impact force is two as previously discussed. However, as this factor increases so does the equivalent static impact force. The dimensionless ratio $(F/w)_{max}$ becomes 4 when $(K_e h/W) = 5$ and even much higher as the height factor continues to increase.

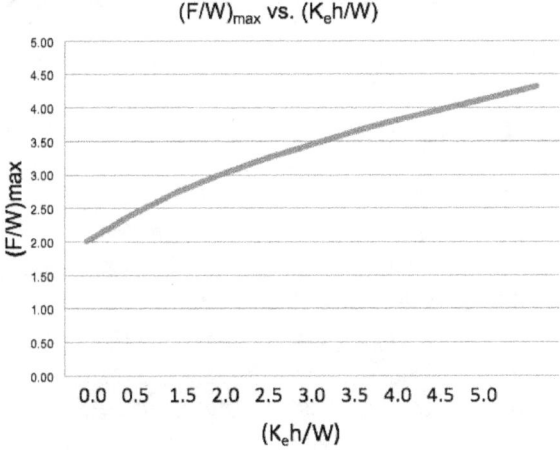

Figure 5 – 8   Variation of $(F/w)_{max}$ with $(k_e h/W)$

Parametric Study

To obtain a better sense of the effects of the system parameters the results of a numerical parametric study are presented in Figure 5-9. Numerical values of F/W are calculated for a drop height of 12 inches together with three sets of k_1 and k_2 values and body weights of 5, 15 and 30 lb. The three values used for k_1 are 2, 20 and 200 lb./in., i.e., low elasticity, a soft spring, to high elasticity, a very stiff spring. For these three values the elasticity of the base is varied from 0.02 lb./in., an extremely soft spring, to 200 lb./in., a quite rigid base.

For a constant value of k_1 the impact force ratio, F/w, increases with increasing values of k_2. These increases become even greater as k_1 is increased. However, for constant values of K_e, like 1.818 and 1.980 the force ratio is the same even though the roles of k_1 and k_2 reverse.

drop height h=12 in.

k1 (lb/in)	k2 (lb/in)	ke (lb/in)	F/w		
			W(5lb)	W(15 lb)	W(30lb)
2	0.02	0.020	2.05	2.02	2.01
2	0.2	0.182	2.37	2.14	2.07
2	2	1.000	3.41	2.61	2.34
2	20	1.818	4.12	2.98	2.57
2	200	1.980	4.24	3.04	2.61
20	0.02	0.020	2.05	2.02	2.01
20	0.2	0.198	2.40	2.15	2.08
20	2	1.818	4.12	2.98	2.57
20	20	10.000	8.00	5.12	4.00
20	200	18.182	10.40	6.49	4.94
200	0.02	0.020	2.05	2.02	2.01
200	0.2	0.200	2.40	2.15	2.08
200	2	1.980	4.24	3.04	2.61
200	20	18.182	10.40	6.49	4.94
200	200	100.000	22.93	13.69	10.00

Figure 5 − 9   Parametric Impact Force Ratio Study

Design for Vibration Isolation and Impact

A hard object dropped onto a soft base is the same as a soft object dropped on to a hard base. Another interesting effect is that increasing the weight at a constant drop distance decreases the impact force ratio for any selection of k_1 and k_2. In some instances, this decrease is small while in others where $k_1 = k_2$ the decrease is significant.

Consider how these results relate to two different physical impact scenarios. In the first case a partially filled glass water pitcher is dropped onto a granite kitchen counter top. The k_1 value of the pitcher would be high, say $k_1 = 200$ lb./in or better as would the k_2 value of the counter top. At the drop height of 12 inches the impact force ratio would be 23 or even higher. The anticipated outcome would be breakage of the pitcher both due to high induced stress and the brittle nature of glass. If the same pitcher is dropped into a box of Styrofoam "peanuts" with a k_2 of say 0.20 the impact force ratio is only 2.37. The pitcher would most probably survive intact. Is it any wonder why Styrofoam packing and air bags are used so extensively for shipping "fragile" objects?

Inelastic Impact

In the second limiting case of inelastic impact bodies collide and remain in contact following the event. The process dissipates energy and physical damage normally takes place. An analysis of the process dynamics makes use of two different interpretations of Newton's 2nd law: the impulse momentum principle and kinetic energy method.

Design for Vibration Isolation and Impact

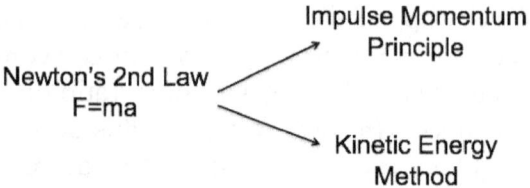

Figure 5 – 10   Two Interpretations of Newton's 2nd Law

Each analysis type possesses advantages and disadvantages but when used together can be a useful tool to determine equivalent static design criteria requirements.

Impulse Momentum Model

As depicted in Figure 5-11 the model is an inelastic body that impacts a fixed wall with an initial velocity $V_1 = V_o$.

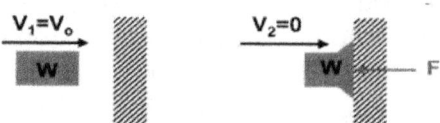

Body impacts fixed object and does not rebound
Apply Newton's 2nd Law

$$-F = ma = \frac{W}{g}\frac{dv}{dt} \quad \text{or} \quad -F\,dt = \frac{W}{g}dv$$

then

$$\int_1^2 F\,dt = \frac{W}{g}\int_2^1 dv = \frac{W}{g}(V_1 - V_2)$$

impulse = change in momentum

Figure 5 – 11   Momentum Impulse Principle

Design for Vibration Isolation and Impact

It then comes to rest, $V_2 = 0$, and deforms under the action of the force F dissipating energy.

The mathematical formulation of this principle is derived from Newton's 2nd law by replacing acceleration with the derivative of velocity by time and integrating both sides of the equation with respect to dt. The final equation governing this formulation appears is the last line in Figure 5-11. The integral of Fdt represents the impulse experienced by the body, W, and the right-side becomes its change in momentum.

The initial and final condition of velocity are now introduced together with an assumed constant force for simplification.

Assume

$$V_1 = V_o, \quad V_2 = 0, \quad F = \text{const.} \quad \text{then}$$

$$\int_1^2 F\,dt = F(t_2 - t_1) =$$

$$\frac{W}{g}\int_2^1 dv = \frac{W}{g}(V_1 - V_2) = \frac{W}{g}V_o$$

or $\quad \dfrac{F}{W}(\Delta t) = \dfrac{V_o}{g} \quad \Rightarrow \quad \dfrac{F}{W} = \dfrac{V_o}{g\Delta t}$

Given the initial velocity it is necessary to specify Δt to determine F/W

Figure 5 – 12   Application to Idealized Model

This application is performed in Figure 5-12 resulting in the final expression that the ratio of the magnitude of the impulse force to the weight of the body is equal to the ratio of the initial velocity to the acceleration of gravity multiplied by the time increment of the event.

The formulation is simple but requires knowledge of how long the process took. It should be noted that F was considered constant during the event to simplify the integration of the impulse. In general, this is not a valid assumption. More on this issue later.

Mass Ratio Effect

Consider how impact response changes if the limiting condition of the fixed base is relaxed. Allow the base to be a finite body that is initially stationary but can experience a velocity following impact. Application of conservation of momentum before and after impact leads to the development in Figure 5-13.

Assume

$V_1 = V_0$, $F = \text{const.}$ then

$$\int_1^2 F\,dt = F\,\Delta t = \frac{W}{g}(V_0 - V_2)$$

Following impact momentum is conserved

$$mV_0 = (m+M)V_2$$

where m = mass of colliding body
M = mass of collided body

set $M = cm$ then

$$V_2 = \frac{1}{1+c} V_0$$

Figure 5 – 13   Effect of Mass Ratio

Design for Vibration Isolation and Impact

This defines the final velocity of the two bodies joined together after impact in terms of the ratio of their masses.

Modified Impulse Force

Substituting this modified expression for V_2 into the general impulse momentum equation in Figure 5-11 results in the inclusion of the mass ratio effect in Figure 5-14.

$$\int_1^2 F\,dt = F\Delta t = \frac{W}{g}(V_0 - V_2)$$

Substutute

$$V_2 = \frac{1}{1+c} V_0 \quad \text{to get}$$

$$\frac{F}{W} = \frac{V_0}{g\Delta t}\left(\frac{c}{1+c}\right)$$

if $c = 1 \quad M = m$

$$\frac{F}{W} = \frac{V_0}{2g\Delta t}$$

Figure 5-14   Mass Ratio Effect on Impact Force

If the mass of the two bodies is equal, $c = 1$ and the impulse force is cut in half of what it would be if c were zero corresponding to a fixed base.

This accounts for why passengers in heavier cars involved in a collision are generally safer than those in lighter cars. The impact force experienced will be less particularly when the impacted car is lighter. The heavier cars are also more generally stoutly designed resulting in less physical damage. However, the passengers themselves are bodies colliding with the interior of the car which is a separate impact problem of its own. Consider how seat belts and air bags impact the passenger impact response in terms of the elastic and inelastic limiting cases analyzed.

Nonlinearity of Mass Ratio Effect

The graph in Figure 5-15 demonstrates vividly the nonlinear behavior of the mass ratio effect.

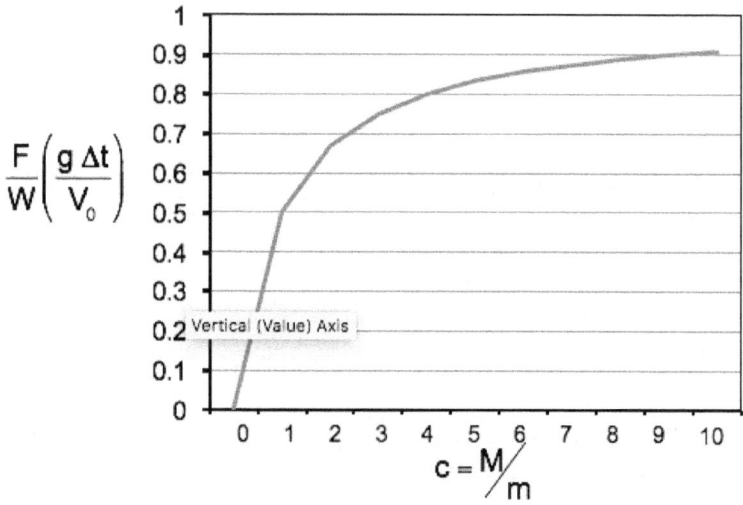

Figure 5-15   Nonlinearity of Mass Ratio Effect

For the range $0<c<4$ the magnitude of the dimensionless impulse force rises rapidly to 0.8. From that point on the increase is asymptotic to 1.0 as c approaches infinity. This corresponding to an immovable base.

Design for Vibration Isolation and Impact

Velocity Ratio Effect

An additional interesting combination of initial conditions prior to impact is when the mass impacted, M, also possesses a velocity V_2 as illustrated in Figure 5-16.

$$V_1 > V_2$$

V_1	V_2		V_f
□	□		◇
m	M		m + M
before impact			after impact

Conservation of momentum

$$mV_1 + MV_2 = (m + M)V_f$$

set $c = M/m$ and solve for V_f

$$V_f = \left(\frac{1}{1+c}\right)(V_1 + cV_2)$$

Figure 5 – 16   Final Velocity, V_f

For impact to take place V_1 must be greater than V2. To include its effect on the force of impact requires applying conservation of momentum to determine the final velocity V_f. This leads to the final equation in Figure 5-16 which also include the mass ratio, c.

Applying Impulse Momentum Principle

The equation for V_f is now substituted for V_2 in the general impulse momentum relationship developed in Figure 5-12. The result is presented in Figure 5-17. In non-dimensional form the coupled effects of mass and velocity ratios on the impact force are clearly identified.

substitute V_f into general momentum impulse equation

$$F \Delta t = \frac{W}{g}(V_1 - V_f) = \frac{W}{g}\left(V_1 - \left[\left(\frac{1}{1+c}\right)(V_1 + c V_2)\right]\right)$$

simplifying results in

$$F \Delta t = \frac{W}{g}\left(\frac{c}{1+c}\right)(V_1 - V_2)$$

in non-dimensional form

$$\frac{Fg}{WV_1} \Delta t = \left(\frac{c}{1+c}\right)(1 - \eta) \quad \text{where} \quad \eta = \frac{V_2}{V_1}$$

Figure 5-17 Mass and Velocity Ratio Effect

Graphical Interpretation

A graphical representation of the last equation in Figure 5-17 for values of mass ratio, M/m, from 0 to 10 and three values of velocity ratio, V_2/V_1, of 0.5, 0.0 and -0.5 is presented in Figure 5-18. It is seen that the impact force rises rapidly with mass ratio in a manner similar to Figure 5-15.   For $V_2/V_1 = 0.0$ the behavior is the same approaching an asymptotic value for the dimensionless impact force of 1.0.   When $V_2/V_1 = 0.5$ the mass M is moving away from the mass m at half its velocity.   The effect is that the impact force is smaller approaching an asymptote of 0.5.   When V_2/V_1 is -.05 the two bodies undergo a head on collision and the impact force increases as expected.   The asymptote in this case is 1.5.

Design for Vibration Isolation and Impact

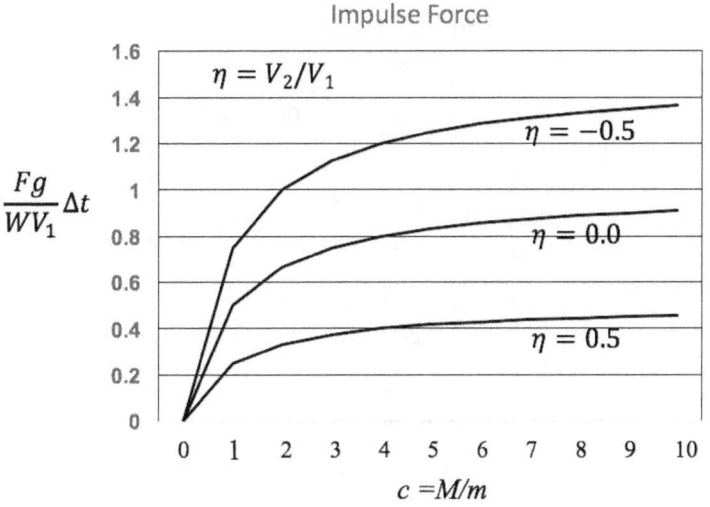

Figure 5-18   Graphical Effects of c and η

Is it any wonder that a head on collision experienced by two automobiles is so destructive?

Kinetic Energy Method

The acceleration in Newton's 2nd Law applied to the impact is now written in the form a = (dv/ds) (ds/dt). Both sides of the equation are integrated with respect to ds over the distance experienced in the event, see Figure 5-19.   The left side becomes the integral of Fds which represents incremental work. The right side becomes the integral of (V) dV which when evaluated becomes a change in kinetic energy.   Thus, the internal work done on the system during impact is reflected in a decrease in its total kinetic energy.

From Newton's 2nd Law

$$-F = ma = \frac{W}{g}a = \frac{W}{g}\frac{dv}{dt} = \frac{W}{g}\frac{dv}{ds}\frac{ds}{dt}$$

then

$$-\int_{s_1}^{s_2} F\,ds = \frac{W}{g}\int_1^2 \frac{ds}{dt}dv = \frac{W}{g}\int_1^2 v\,dv$$

$$\int_{s_1}^{s_2} F\,ds = \frac{W}{2g}\left(V_1^2 - V_2^2\right)$$

Work done on body =
change in it's kinetic energy

Figure 5-19   Kinetic Energy Method

Limiting Case Result

The limiting case conditions from Figure 5-12, a body impacting a rigid base, are now applied to the kinetic energy formulation. The development is presented in Figure 5-20. The result is the dimensionless impulse force is equal to V_o^2 divided by 2g times Δs, the distance the impacting body moves as a consequence of the event. As in the impulse momentum principle it is again necessary to know something besides the velocity and mass of the colliding body to calculate an impact force. The distance moved following initial contact, Δs, must be specified.

The force, F, is again assumed to be constant to simplify the integration since its variation with s is unknown.

Design for Vibration Isolation and Impact

Assume

$V_1 = V_o$, $V_2 = 0$, $F = \text{const.}$ then

$$\int_1^2 F\,ds = F\Delta s =$$

$$\frac{W}{g}\int_1^2 v\,dv = \frac{W}{2g}(V_1^2 - V_2^2) = \frac{W}{2g}V_o^2$$

or $\quad \dfrac{F}{W}(\Delta s) = \dfrac{V_o^2}{2g} \;\Rightarrow\; \dfrac{F}{W} = \dfrac{V_o^2}{2g\,\Delta s}$

Given the initial velocity it is now necessary to specify Δs to determine F/W

Figure 5 – 20   Limiting Case Impact Result

The effect of mass ratio where the base is replaced by a second body of finite mass is identical to that developed for the impulse moment approach.

Example Design Application

A deformable automobile bumper system is to be designed to deflect 6 inches when the car impacts a rigid barrier at 5 mph. What magnitude of applied static loading should the bumper be designed for. The car weighs 2500 lbs.

Application of the kinetic energy method solution for impact works best in this instance since the allowable deformation is specified. With $V_o = 73.3$ ft./sec (5 mph) and $\Delta s = 6$ inches (0.5 ft.) the impulse force ratio is calculated to be 1.67 in Figure 5-21. This results in an equivalent static design force of 4175 lbs.

73

Design for Vibration Isolation and Impact

$$V_o = 5 \text{ mph} = 5 \text{ mph}\left(\frac{88 \text{ ft/sec}}{60 \text{ mph}}\right) = 7.33 \text{ ft/sec}$$

$$\Delta s = 6 \text{ inches} = \frac{6}{12}\text{ft} = 0.5 \text{ ft}$$

$$\frac{F}{W} = \frac{1}{\Delta s}\left(\frac{V_o^2}{2g}\right) = \left(\frac{1}{0.5}\right)\left(\frac{7.33^2}{2 \times 32.2}\right) = 1.67$$

with W = 2500 lb   then   F = 4`175 lb

Figure 5 – 21   Equivalent Static Design Force

Variable Impulse Force

A more realistic impulse force assumption is that F will increase from zero to its maximum value over the allowable 6 inches of deformation. This is expressed analytically in Figure 5-22.

$$F = \left(\frac{F_{max}}{S_{max}}\right)s$$

Now apply kinetic energy principle

$$\int F \, ds = \left(\frac{F_{max}}{S_{max}}\right)\int_0^{S_{max}} s \, ds = \left(\frac{F_{max}}{S_{max}}\right)\left(\frac{s_{max}^2}{2}\right) = \left(\frac{F_{max} S_{max}}{2}\right) = \left(\frac{W}{2g}\right)V_o^2$$

$$F_{max} = \frac{W V_o^2}{S_{max} \, g} = \frac{(2500)(7.33)^2}{(0.5)(32.2)} = 8343 \text{ lb}$$

Figure 5 – 22   Variable Impact Force Result

Performing the indicated integration of the linear behavior of F with respect to s on the left side of the equation gives the final analytic result in Figure 5-22.

Design for Vibration Isolation and Impact

Substituting the numerical values of the problem parameters into the F_{max} relation gives a new equivalent static design load of 8343 lbs. This is twice the value calculated for a constant force and should be more realistic.

Time of the Event

With estimates of the magnitude of the impact force established the impulse momentum principle can be applied to determine how long the event takes. With the impact force assumed constant the time of the event is calculated as 0.14 sec in Figure 5-23.

$$\frac{F}{W} \Delta t = \frac{V_o}{g}$$

$$\Delta t = \left(\frac{W}{F}\right)\left(\frac{V_o}{g}\right) = \left(\frac{1}{1.67}\right)\left(\frac{7.33}{32.2}\right) \text{ sec}$$

$\Delta t = 0.14$ sec   Is this reasonable?

Figure 5-23   Event Time Duration

If the force is assumed to increase linearly with time the duration of the event will be 0.07 sec. This should be more realistic. Readers should satisfy themselves that this value is correct.

Design for Vibration Isolation and Impact

Chapter 6 – Heavy Springs

Introduction

It was assumed in all the previous analyses that the support springs had no mass. The effect of the weight of the spring will now be taken into account. Specifically, the change in the natural frequency of a single degree system due to the additional weight of its support spring will be determined.

Idealized model

The model used to examine the effect of spring weight is shown in Figure 6 - 1. The coordinate y will designate any location along the length of the spring. An incremental length of spring will be designated as dy, $\overline{w}$ is the weight of the spring per unit length and L is its effective length.

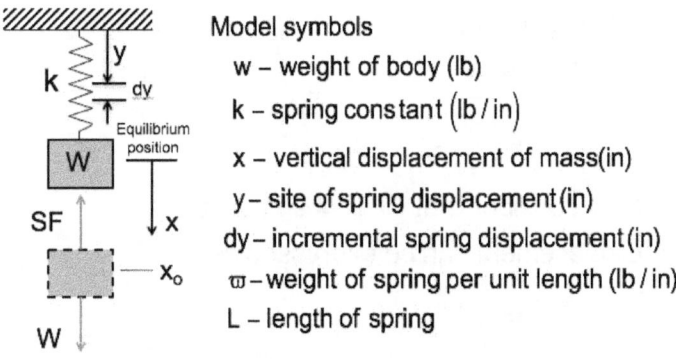

Model symbols

w – weight of body (lb)

k – spring constant (lb/in)

x – vertical displacement of mass (in)

y – site of spring displacement (in)

dy – incremental spring displacement (in)

$\overline{w}$ – weight of spring per unit length (lb/in)

L – length of spring

Figure 6 – 1 Idealized Single Degree Model

Design for Vibration Isolation and Impact

As in previous analyses the displacements of the weight W will be measured from it static equilibrium position.

Harmonic Response

In Figure 6–2 it is assumed that the vertical motion of the displaced weight W will be harmonic. This is expressed as $x = x_o \sin \omega t$. If $\bar{w}$ is much less than W, the weight of the spring can be neglected relative to the weight W, and from previous analysis the natural angular frequency is $\sqrt{kg/W}$. Now consider the case where the weight of the spring can't be neglected. It is assumed that the spring will follow the harmonic motion of the weight W. This is expressed as $s = (y/L) x_o \sin \omega t$. When y is zero there is no motion in the spring where it is attached to the fixed base. When y is equal to L the spring has the same displacement as the weight W to which it is attached. In between its displacement is linearly distributed.

Assume harmonic system behavior

$x = x_o \sin \omega t$

if $\varpi L \leq W$

then $\omega = \sqrt{\dfrac{kg}{W}}$

Now consider that $\varpi L \not\leq W$ and spring displacement can be expressed as

$s = \dfrac{y}{L} x_o \sin \omega t$

Figure 6 – 2   Harmonic Response

Design for Vibration Isolation and Impact

Kinetic Energy of Spring

In preparation for the application of the potential and kinetic energy principle to this assumed conservative system it is first necessary to determine the kinetic energy of the moving spring. The spring velocity is obtained by differentiating its displacement with respect to time as shown in Figure 6–3. This results in $ds/dt = (y/L) \omega x_o \cos \omega t$. It follows then that the maximum velocity is given simply by $(y/L) \omega x_o$ when $\omega t = \pi$ and $\cos \omega t = 1$. The incremental maximum kinetic energy of the spring $d(T_{max})$ is just $1/2\, d(mv^2)$. But dm can be replaced by $\overline{w}/g\, dy$ and v^2 is simply $(y/L\, \omega x_o)^2$. This incremental energy expression is integrated over the length of the spring to give the total maximum kinetic energy of the spring as
$\frac{1}{2}(\overline{w}L/3g)(\omega x_o)^2$.

Spring Velocity
$$\frac{ds}{dt} = \frac{y}{L}\omega x_o \cos \omega t \Rightarrow \left[\frac{ds}{dt}\right]_{max} = \frac{y}{L}\omega x_o$$

Maximum kinetic energy (spring)
$$dT_{max\,s} = \frac{1}{2}d(mv^2) = \frac{\overline{w}}{2g}dy\left(\frac{y}{L}\omega x_o\right)^2$$

$$T_{max\,s} = \frac{\overline{w}}{2g}\left(\frac{\omega x_o}{L}\right)^2 \int_0^L y^2 dy = \frac{1}{2}\left(\frac{\overline{w}L}{3g}\right)(\omega x_o)^2$$

Figure 6 – 3   *Kinetic Energy of Spring*

Design for Vibration Isolation and Impact

Total Maximum Kinetic Energy

The maximum kinetic energy of the entire system is obtained by adding the maximum kinetic energy of the weight W to the maximum kinetic energy of the spring. For the weight $T_{maxW} = \frac{1}{2}(W/g)(\omega x_o)^2$. The final equation in Figure 6-4 is the total system maximum kinetic energy.

$$T_{max\,W} + T_{max\,s} = \frac{1}{2}\left(\frac{W}{g}\right)(\omega x_o)^2 + \frac{1}{2}\left(\frac{\varpi L}{3g}\right)(\omega x_o)^2$$

$$T_{max\,total} = \frac{1}{2}\left(\frac{W + \frac{\varpi L}{3}}{g}\right)(\omega x_o)^2$$

Figure 6 – 4   Total System Maximum Kinetic Energy

Total System Potential Energy

To apply the energy principle the maximum potential energy of the system is also required. This is developed in Figure 6-5 by integrating the incremental strain energy over an incremental extension, dx, of the spring over its total displacement x_o to give $\frac{1}{2} kx_o^2$.

$$dU = (kx)dx$$

$$U_{max} = \int_o^{x_o}(kx)dx$$

$$U_{max} = \frac{1}{2}kx_o^2$$

Figure 6 – 5   System Maximum Potential Energy

Design for Vibration Isolation and Impact

Applying Energy Principle

In a conservative system where no energy is dissipated the total maximum kinetic energy is equal to the total maximum potential energy. Setting these two terms equal to one another in Figure 6-6 and solving for ω^2 gives the quantity kg over $W + \frac{\overline{w}L}{3}$. The frequency of the system becomes:

$$f = \frac{1}{2\pi} \sqrt{\frac{kg}{W + \overline{w}L/3}}.$$

The natural frequency of the system taking into account the mass of the spring is the same as the natural frequency without the spring mass with one-third the weight of the spring added to the weight W. This can be taken as a general rule of thumb irrespective of how the elastic support of W is provided.

Set

$$T_{max\ total} = U_{max}$$

$$\frac{1}{2} \left(\frac{W + \frac{\overline{w}L}{3}}{g} \right) (\omega x_o)^2 = \frac{1}{2} K x_o^2$$

Solve for ω^2

$$\omega^2 = \left(\frac{Kg}{W + \frac{\overline{w}L}{3}} \right) \Rightarrow f = \frac{\omega}{2\pi} = \frac{1}{2\pi} \sqrt{\frac{Kg}{W + \frac{\overline{w}L}{3}}}$$

Figure 6 – 6   Applying Energy Principle

Design for Vibration Isolation and Impact

Example Problem

A 50-lb. weight is supported vertically by a 12-foot long half-inch diameter steel rod. Determine the natural frequency of the weight both with and without taking into account the weight of the supporting rod. Without the weight of the rod the natural angular frequency is given by the $\sqrt{kg/W}$. The numerical value of the k needs to be determined first. The spring constant, k, for a rod is given by its cross-sectional area, A, times the modulus of the material, E, divided by it length L.

Substituting the appropriate number from the problem into this expression with the modulus of steel taken as 30 x 10^6 lbs./inch^2 results in k = 408 x 10^2 lbs./inch in Figure 6-7. Substituting this into the expression for ω gives a value of 501 radians/second or a final frequency of 89.3 hertz.

Without mass of spring

$$\omega = \sqrt{\frac{kg}{W}}$$

Spring constant

$$k = \frac{AE}{L} = \frac{\left(\pi(1/4)^2\right)(in)^2\left(30 \times 10^6\right)(lb/in^2)}{144(in)} = 408 \times 10^2 \, (lb/in)$$

$$\omega = 10\sqrt{\frac{(408)386}{50}} = 501 (rad/sec) \Rightarrow f = \frac{\omega}{2\pi} = 89.3 \, hertz$$

Figure 6 – 7   Frequency without Spring Weight

Design for Vibration Isolation and Impact

Include Spring Weight

The weight of the support rod is given by its cross section area A multiplied by its length L times its weight density per cubic inch, ρ. Using the weight density of steel as 460 lb./ ft.^{3} the weight of the rod is calculated to be 7.5 lbs. in Figure 6-8. Adding one third of this value of 2.5 pounds to the weight of the supported weight gives a corrected value of the frequency of 87.3 hertz.

Include weight of spring

$$W_s = A L \rho = \left(\pi(1/4)^2\right)(\text{in})^2 \times 144(\text{in}) \times \frac{460(\text{lb}/\text{ft}^3)}{1728(\text{in}^3/\text{ft}^3)}$$

$$W_s = 7.5 \text{ lb}$$

$$\omega = \sqrt{\frac{Kg}{W + \frac{W_s}{3}}} = 10\sqrt{\frac{(408)386}{52.5}} = 54.3 \text{ (rad/sec)}$$

$$f_s = \frac{\omega}{2\pi} = \frac{54.3}{2(3.14)} = 87.3 \text{ hertz}$$

$$\%\text{difference} = \left(\frac{89.3 - 87.3}{89.3}\right) 100 = 2.2 \%$$

Figure 6 – 8   Including Spring Weight

The percentage difference from the uncorrected value is 2.2 %. In this instance neglecting the weight of the rod did not significantly change the frequency and demonstrates why in most cases the weight of the spring can be neglected. However, a good rule of thumb has been determined for including the weight of the elastic support where it might be important.

Limiting Case

This approximation can also be used to determine the natural frequency of the rod itself by setting W =0 in the previous calculation. This results in a frequency of the spring, oscillating vertically without the attached weight, of 400 hertz. Even though this is 4.5 times the frequency of the combined system its effect on the system frequency is only 2.2%. More importantly, how accurate is the result of 400 cps since this does represent an approximation. A detailed analysis of wave propagation in the spring is required to answer this question.

Idealized Wave Mechanics Model

The idealized physical model used to analyze one dimensional wave propagation simulating free spring oscillation is illustrated in Figure 6-9. It consists of a longitudinal bar fastened at its upper end and free at its bottom. Its weight is uniformly distributed, all damping is neglected and only movement in the x direction is permitted. As the bar oscillates an element dx undergoes a displacement u from its original x position and is subjected to a compression or extension depending on how a longitudinal displacement wave propagates. This behavior is governed by a second order partial differential equation in which u is a function of both position, x, and time, t. (The development of this equation and its solution is beyond the mathematics of this monograph. However, an overview of its results and consequences are presented for completeness.)

Design for Vibration Isolation and Impact

$$\frac{\partial^2 u}{\partial x^2} = \frac{1}{c^2}\frac{\partial^2 u}{\partial t^2} \quad \text{where} \quad c^2 = \frac{Eg}{\rho}$$

general solution

$$u = F(x)(C_1 \sin \omega t + C_2 \cos \omega t)$$

where

$$F(x) = A \sin\frac{\omega x}{c} + B \cos\frac{\omega x}{c}$$

Figure 6 – 9   Idealized Wave Mechanics Model

Governing Equation and Solution

The equation that governs the wave propagation is a second order partial differential equation in the displacement, u, as a function of position, x, and time, t, with $c^2 = Eg/\rho$, in Figure 6 – 9. Its solution can be expressed as a function $F(x)$ multiplied by a harmonic in ωt. The function $F(x)$ is also a harmonic whose argument is $\omega x/c$.

Boundary Conditions

The constants A and B in F(x) are dependent on the boundary conditions on displacement, u, applicable to the model. In this example there is no strain, $du/dx = 0$, at the free end and no displacement, $u=0$, at the fixed end. These are applied in Figure 6-10.

The two resulting equations can only be satisfied ii $A=0$ and $cos(\omega l/c) = 0$. This implies that $(\omega l/c) = (\pi/2)$.

Design for Vibration Isolation and Impact

boundary conditions

$\frac{\partial u}{\partial x} = 0$ at $x = 0$, $u = 0$ at $x = l$

then

$\left(\frac{\partial u}{\partial x}\right)_{x=0} = A\frac{\omega}{c}(C_1 \sin \omega t + C_2 \cos \omega t) = 0$

$u_{x=l} = \left(A \sin \frac{\omega l}{c} + B \cos \frac{\omega l}{c}\right)(C_1 \sin \omega t + C_2 \cos \omega t) = 0$

so that

$A = 0$, $\cos \frac{\omega l}{c} = 0$ and $\frac{\omega l}{c} = \frac{\pi}{2}$

Figure 6 – 10   Applying Boundary Conditions

Frequency Comparison

A comparison of the energy frequency and wave equation frequency is shown in Figure 6-11. From the last requirement in Figure 6-10, the wave frequency is:

$$\omega = \frac{1.57}{l}\sqrt{\frac{Eg}{\rho}}.$$

Expressing the energy principle frequency in the same variables as the wave frequency results in

$$\omega = \frac{1.73}{l}\sqrt{\frac{Eg}{\rho}}.$$

This is about 10 % higher than what is considered a more accurate solution.

Design for Vibration Isolation and Impact

Wave frequency

$$\frac{\omega l}{c} = \frac{\pi}{2} \quad \rightarrow \quad \omega = c\frac{\pi}{2l}$$

$$\omega = \frac{\pi}{2l}\sqrt{\frac{Eg}{\rho}} \quad \rightarrow \quad \omega = \frac{1.57}{l}\sqrt{\frac{Eg}{\rho}}$$

Energy principle frequency

$$\omega = \sqrt{\frac{kg}{w/3}} \quad \text{but} \quad k = AE/l \quad \text{and} \quad w = Al/\rho$$

so that

$$\omega = \frac{1.73}{l}\sqrt{\frac{Eg}{\rho}} \quad \sim 10\% \text{ higher}$$

Figure 6 – 11   Frequency Comparison

Even in this extreme limiting case for the energy principle approximation the results are not excessively different. This provides even greater confidence in the value of this simple approximation for including the mass effects of the spring support where necessary.

Design for Vibration Isolation and Impact

Design for Vibration Isolation and Impact

Spring Constants

A sampling of equivalent spring constants is presented for rods in extension, shafts in torsion and beams in bending that can be used as elastic supports. These formulas are expressed in terms of cross section geometries, material properties and length.

Rods in Extension

The extensional spring constant for two examples of straight rods are presented in Figure 1. One rod is of constant cross section area, A, while the second is stepped with each step being of constant cross section, A_1 and A_2. The modulus of elasticity of the material is designated E. Each carries the same weight, W. The deflections are derived from strength of materials models assuming linear behavior.

Straight Rod
$$\delta = \frac{Wl}{AE} \quad k = \frac{AE}{l}$$

Stepped Rod
$$\delta = W\left(\frac{l_1}{A_1 E_1} + \frac{l_2}{A_2 E_2}\right) = W\left(\frac{1}{k_1} + \frac{1}{k_2}\right)$$

$$k = \frac{1}{\left(\frac{1}{k_1} + \frac{1}{k_2}\right)} = \left(\frac{k_1 k_2}{k_1 + k_2}\right)$$

Figure 1 Spring Constants – Rods in Extension

Design for Vibration Isolation and Impact

Shafts in Torsion

Spring constants for a constant cross section and stepped cross section shaft in torsion are presented in Figure 2. The cross-section property is the polar moment of inertia, J. The material property, G, is the torsional modulus of elasticity. It is observed that the format of these formulas is similar to those for a rod in extension.

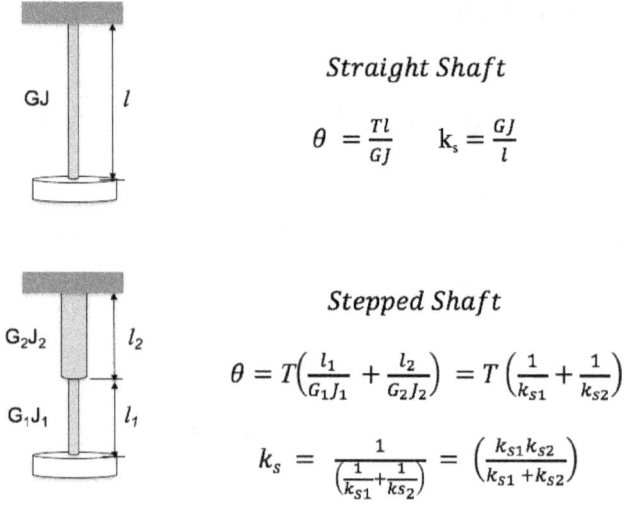

Figure 2   Spring Constants – Shafts in Torsion

Beams in Bending

Spring constants for three beam examples are presented in Figure 3. These include a cantilever beam with the load a the end, a simply supported beam with the load in the middle and a beam built in at both ends with the load in the center.

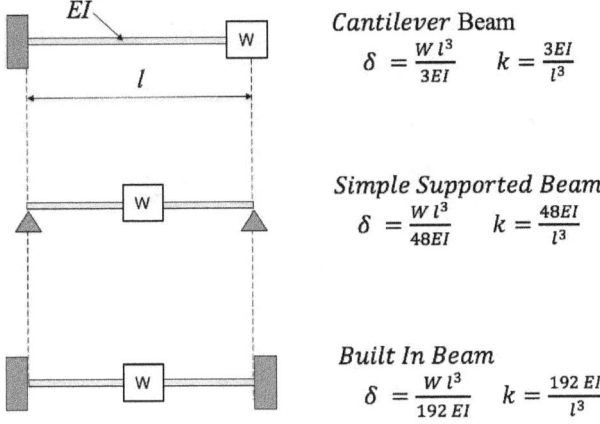

Figure 3   Spring Constants – Beam in Bending

The units of all cross-section dimensions, material properties and length parameters must be chosen appropriately to provide the units desired for the spring constants.

Design for Vibration Isolation and Impact

About the Author

Carl F. Zorowski is R. J. Reynolds Professor Emeritus of Mechanical and Aerospace Engineering at North Carolina State University. He attended Carnegie Mellon University receiving his doctorate in 1956. He taught at CMU until 1962 before joining North Carolina State University.

His academic career at NCSU included teaching, research and administration serving as Head of the Mechanical and Aerospace Engineering Department and Associate Dean for Academic Affairs in the College of Engineering. He was also cofounder and Director of the Integrated Manufacturing Systems Engineering Institute, an early interdisciplinary master's practice program at NCSU.

His passion is classroom instruction, course content development and delivery methods that emphasize mathematical modeling and design performance analysis of mechanical systems. In 1993, he directed a five-year NSF funded Engineering Education Coalitions Program dedicated to revitalizing undergraduate engineering instruction involving eight southeastern universities. Following retirement in 1997 he has offered graduate level courses in Mechanical Design Engineering in the College's distance education master's degree program.

He is a fellow of the American Society of Engineering Education and the American Society of Mechanical Engineers.

Design for Vibration Isolation and Impact

Other monographs by the author

Design of Mechanical Power Transmissions
Design for Assembly
Design for Static Mechanical Strength
Design for Mechanical Fatigue
Design for Bending, Torsion and Buckling
Design for Deflection
Solving Engineering Problems

Design for Vibration Isolation and Impact

www.ingramcontent.com/pod-product-compliance
Lightning Source LLC
Chambersburg PA
CBHW070303230526
45470CB00002B/700